In God's Countries

By Bil Gilbert

Foreword by Robert W. Creamer

IN GOD'S COUNTRIES

University of Nebraska Press

Lincoln and London

Acknowledgments for the
use of copyrighted material
appear on page viii.

Copyright 1984 by the
University of Nebraska Press
All rights reserved
Manufactured in the United
States of America

The paper in this book meets
the guidelines for permanence
and durability of the Committee
on Production Guidelines for
Book Longevity of the Council
on Library Resources.

Library of Congress
Cataloging in Publication Data

Gilbert, Bil
In God's countries.
1. Natural history –
Addresses, essays, lectures.
I. Title.
QH81.G48   1984
508   83-23384
ISBN 0-8032-2117-7
(alk. paper)

To Sam Walmer

It has been written that he is "loud, iconoclastic, disrespectful, observant and curious, all of which makes him a stimulating traveling companion." That he has been for many of these—and other—improbable ventures.

Contents

        Acknowledgments, viii

        Foreword, ix

Part I Other Bloods, 1

        The Devil in Tasmania, 3

        The Second Keeper, 22

        The Missouri Kid, 42

        An Eyeshine of Ferrets, 59

        Turn of a Century, 77

        Old Fish, 91

Part II In God's Countries, 101

        The Rites of Autumn, 103

        Baffinland, 123

        They Crawl by Night, 141

        Trailing, 157

        Going Under, 172

        Journey into Spring, 185

ACKNOWLEDGMENTS

The following articles are reprinted courtesy of *Sports Illustrated* from the issues indicated:
"The Devil in Tasmania," originally published as "Nasty Little Devil," October 5, 1981. © 1981 Time Inc.
"The Second Keeper," originally published as "Keeper of Something Unique," November 5, 1979. © 1979 Time Inc.
"The Missouri Kid," originally published as "Goin' South," June 11, 1979. © 1979 Time Inc.
"An Eyeshine of Ferrets," originally published as "Missing and Presumed Extinct," October 13, 1980. © 1980 Time Inc.
"Turn of a Century," originally published as "Final Rendezvous in the Seven Mountains," December 11, 1972. © 1972 Time Inc.
"The Rites of Autumn," November 27, 1978. © 1978 Time Inc.
"Baffinland," originally published as "Cold Place for a Walk," February 12, 1979. © 1979 Time Inc.
"They Crawl by Night," August 27, 1979. © 1979 Time Inc.
"Trailing," originally published as "Range of Diversity," January 14, 1980. © 1980 Time Inc.
"Going Under," originally published as "Journey to the Center of the Earth," May 20, 1974. © 1974 Time Inc.
"Journey into Spring," May 8, 1978. © 1978 Time Inc.

"Old Fish" was originally published as "End of a Long Journey for Spoonbill Cat" in *Audubon,* March, 1981. Copyright 1981 by Bil Gilbert.

Foreword

When Bil Gilbert spoke to me about doing the introduction to this book he said, "Would you like to write the warning?" I was amused by his use of the word, but I think now that—facetiously or not—he applied it properly.

Be warned: these pages may become addictive.

I know that when my daughter came home to spend a few days with us before her marriage this spring, she noticed the pile of Gilbert manuscripts that I was rereading before beginning to write this. She has never met Bil—both tend to be peripatetic, and their irregular courses have not yet coincided—but she had read stories of his in *Audubon* and *Smithsonian* and *Sports Illustrated*. Although she had a million things to do before the wedding, she sat down "just for a minute" to read one of these pieces and remained there for hours, reading all of them. Gilbert does that to you.

Yet he's not the sort of writer who jumps off the page and grabs you by the throat with zingy, melodramatic copy. He doesn't seem to worry, as some writers do, that if he doesn't catch you with his first sentence, or his first paragraph, he may lose your attention forever. Not for him is the no-holds-barred opener: "Our small plane barely sputtered its way over the jagged knife-edge of the ridge. Alvin 'Biff' Sturdley, our veteran bush pilot, gritted his teeth and miraculously guided the craft to a delicate landing on a tiny patch of clearing next to the alligator-infested Ruratura River."

Gilbert is different. He likes to back and fill before getting to the meat of a story. He kind of sidles around it, walking back and forth and talking over his shoulder before he settles down. What he's doing is making you familiar with the general subject at hand before he plunges into the particulars and pulls you after him.

This is partly a product of habit, I suppose, developed during long years of setting off on journeys, adventures, expeditions, explorations, walks, hikes, canoe trips, biological field trips, and visits to editors and publishers. He's used to getting himself ready before he sets out, sorting out what's necessary, leaving behind the nonessentials. He wants his readers to be prepared, too, so he spends a few paragraphs doing just that. "By way of preliminary" and "by way of explanation" seem to me to be typical Gilbertian phrases, although I couldn't swear that either actually appears in any of the choice, high-quality, grade-A Gilbert stories that make up this book. He likes to establish the reason for a story, the premise, the background, and he makes sure that you have the picture—that you are prepared, in other words—before he gets down to the pith.

In "The Missouri Kid," for example, his marvelous account of a very individualistic moose, he spends three good-sized paragraphs talking about graduate students, range maps, and the surface of the moon before he deftly, with perfect timing, gets around to his first use of the word "moose." In "Going Under," which is about exploring caves, he doesn't even mention caves until we're 600 words into the story, yet his skillful survey of the topography and flora of the Appalachian countryside leads us to the entrance to the caves as eager for them as though we had been along on the trip. His account of following animal trails in the Huachuca Mountains of southern Arizona, "Trailing," begins with a discussion of cottontail tracks on suburban lawns and cat prints in the back alleys of New York City's west side.

Beyond habit, this roundabout approach to a story is probably a result of politeness. You wouldn't dream of turning to a stranger on a bus and beginning a long story about something that happened to you without doing a little briefing first, putting your listener at ease and setting him up so that he's better able to understand and appreciate what you're talking about. So with Gilbert.

Finally, there is his predilection for instructing. Although he has been a full-time free-lance writer most of his adult life, he has from time to time served as both a teacher and an athletic coach, professions in which he has had the responsibility of helping people to absorb knowledge. To teach or coach successfully (and he was very successful), you have to prepare your charges before you can slip them the doses of information you have readied for them.

Gilbert, in a sense, is always teaching and coaching. ("If you yell at me, I'll yell back," his daughter Lyn warns him in "Journey into Spring" as they begin a canoe trip down the West Branch of the Susquehanna River. "You'll be all right," Gilbert tells her. "Anyway, I never did yell much—just coached forcefully.") He's always discovering things—in his reading, in his research, in his travels—and he feels an imperative to pass them along in his writing.

Yet he avoids pedantry, a fault many teachers (and coaches) fall into at one time or another. Gilbert has fairly strong opinions, but he has the confident man's ability to hold them back and not only listen to but understand an opposite point of view, however much he disagrees with it. He doesn't try to beat information into your head. He shows it to you, as he might call your attention to a raptor perched on a dead branch of a tree near a river or to the fact that you, the non-hunter, non-outdoorsman, get a thrill out of seeing rabbit tracks on your side lawn after a light snow.

Except that Gilbert doesn't challenge you with a word as demanding as "thrill." He low-keys it. In "Trailing" he says of rabbit tracks on the lawn, "I have yet to meet anyone . . . who is bored by the discovery," which tends to induce voluntary agreement. Nor does he patronize the reader; in the same piece, after talking about following those cat tracks in Manhattan, he says, "Because I like the Yukon Valley better than West 80th Street, I found cat-trailing inferior to pursuing wolves, but it was nonetheless genuine trailing, and a better way to kill a morning than hanging around Zabar's." (Zabar's, as an omnivorous devourer of topographical information like Gilbert knows, is a chi-chi delicatessen on Manhattan's upper west side.)

Beyond being teacher, coach, instructor, and guide, Gilbert is above all else a writer, and a fine one. When he and his friend Sam Walmer climb an icy ridge of Cumberland Peninsula on Baffin Island in "Baffinland," you don't get an ego-oriented account of a wrestle with nature or a painfully detailed lecture on glaciation, but a story about people—and a fairly dramatic story at that, in the sense that the personalities in it move and develop, act and react. There is a climax and an anticlimax, all without exaggeration or hyping-up of the events of the trip. Some passages are poetic in their impact. The most vivid description of an elephant I've ever read is in the Baffin Island piece. It is not an aside, or a clever paradox or a writer's conceit, but an intrinsic part of Gilbert's perception of Cumberland Peninsula. I thought of Hemingway's "Hills Like White Elephants" when I read Gilbert's controlled but highly emotional reaction to a glacier close up.

While Bil, a naturalist by education and training, is particular about qualifying his statements (a naturalist is seldom absolutely *sure* of anything; conclusions must always be open to further testing), he can whip out flat, uncompromising word sketches that are as cutting as a David Levine caricature. His picture of the unsavory beast called the Tasmanian devil in "The Devil in Tasmania" is a masterpiece of objective vituperation. Part of it goes: " . . . the devil is so hinged as to be able to open its mouth very wide, and it does this often, being habitually slack-jawed and gapish. Also, it's a steady drooler. The devil has prominent, almost hairless, batlike ears, small mean eyes, the long, coarse whiskers of a big rat and a piggish, dripping nose. Its body is lumpish, overlaid in maturity with heavy layers of fat. Its legs are bandy, with the rear ones giving the impression of being disproportionately long, lending a jacked-up appearance. . . . Its ordinary pace is a shamble. When it needs to move more rapidly, it lurches. . . . As with people, some animals—English bulldogs come to mind—are able to overcome ugliness with charming personalities. The Tasmanian devil isn't among these."

More amiably, Gilbert describes other Tasmanian wildlife: "Wombats are about the size and shape of furry medicine balls with facial expressions like those of elderly academics" and "Bandicoots are rabbity creatures with very long pointed, bewhiskered, inquisitive faces. When pursued they hop about as if on pogo sticks, and they probably retire to C. S. Lewis' Narnia during the day."

Although Gilbert writes with affection, or at least appreciation, of animals, he avoids anthropomorphizing them. In his account of the splendidly aberrant moose mentioned earlier, which took off from its normal territory in northern Minnesota and walked through farmland

and towns and at least one good-sized city to a point more than 600 miles south of the southern limit of its supposed range, he achieves something extraordinary. Without falling into the anthropomorphic trap, he humanizes the moose. I should say he "animalizes" him, to use a parallel to the ordinary meaning of humanize—that is, he makes him credible and available to understanding. He establishes that this appealing animal has blithely disregarded the rigid laws of behavior established for mooses by those who are not mooses. I use that plural not to be arch but to emphasize what Gilbert emphasizes in his story: that this moose was not one of a matched set of undeviating things called moose but was truly an individual, as are most animals—and people—when you get to know them.

The moose's well-publicized journey south through Minnesota and Iowa into central Missouri gave many skillful journalists the inspiration for hilariously funny pieces in print and on TV and radio. Gilbert has fun, too, as you'll find when you read the story, but at the same time he treats the moose's journey seriously. I don't mean solemnly, with scientific measurements and ponderous conclusions. I just mean that while Gilbert enjoys the wanderings of the moose as much as anyone, he also explores the possible reasons why the moose did what he did, and you come out of his story with a lot more than a momentary chuckle about an oddball animal.

Gilbert's humor, and there's plenty of it strewn through these pages, is never trivial, not even in "They Crawl by Night," his visit to an earnest earthworm entrepreneur named Ray Edwards, who wrote a paperback book on the subject called *The Nightcrawler Manual*. Worms. Oh, boy. Riper material than the moose for a kicky newspaper feature or a 30-second spot on the evening news—a couple of gags, a few quick quotes, and on to the next item. Gilbert obviously is amused by Edwards and his attachment (so to speak) to worms, and his report of a nocturnal worm hunt with Edwards is pretty funny. But he treats both Edwards and the worms with basic respect, and before you know it you've learned a surprising lot about the little wrigglers—good, valuable information, too. Did you know, for instance, that Darwin rated worms near the top among creatures that have played a particularly important role in the history of the world?

As an environmentalist, Gilbert writes perceptively about depleted or vanishing species, from the black-footed ferret ("Missing and Presumed Extinct") to the paddlefish ("Old Fish"), but he writes just as sensitively about human beings involved in animals' lives. I don't mean just those who are dedicated to trying to save a species, like desperate, middle-aged Jack Lynch trying to preserve the rare buffalo wolf in "The Second Keeper," but those who have done some of the destroying, too, like the beleaguered frontiersman in Pennsylvania at the end of the 1700s who slaughtered eastern buffalo herds to the last animal as a matter of survival ("Turn of a Century"). Gilbert's grasp of history and social necessities is too strong to allow him to condemn such behavior out of hand, or even that of modern-day Pennsylvanians who kill as many as 150,000 deer in a single hunting season, more than half of them the first day ("Rites of Autumn"). He doesn't preach about these things; he just tells you about them, in gripping detail. He's a masterful storyteller, with a true gift for bringing a narrative to a gripping conclusion.

A case in point is his wry, fairly short essay on the paddlefish, a remarkable freshwater creature of the Mississippi Basin that survived practically unchanged and in great numbers through the vicissitudes of time for more than 300 million years, is now threatened with extinction because of changes in the river system (primarily dams and lakes instead of running streams), and is being kept alive in good numbers only through the efforts of conservation officials who have the support of sportsmen who like to catch the fish.

Gilbert ends his paddlefish story: "That we have figured out how to do these things to and for such an ancient creature is a credit to our intelligence. That we want to do them speaks well for our compassion—or at least says something about how well we regard sport. That we need to do these things—that the paddlefish, her attendants, and the rest of us have met for these purposes at Blind Pony Hatchery—is unspeakably sad."

The climaxes of the stories about the Tasmanian devil and the animal trails in the Huachuca Mountains and the mass deer hunt reach an intensity that is extraordinary. His understanding of the motives of men like the field biologists who persist in their almost fruitless search for the all but extinct ferret, the frontiersmen who ruthlessly slaughtered the marauding buffalo, and Lynch, who left his wife and his job and has literally given his life, or the living of it, for his wolves, is broad and extensive. What Gilbert does best is make you think.

He makes you enjoy, too. His style isn't actually breezy, but there's a pleasant irreverence in it that keeps it from ever getting heavy. Before he and Lyn begin their canoe trip down the sometimes breathtakingly lovely West Branch, he pauses to note that the stream rises "from a number of seeps in a swampy bowl overlooked by several junk-food drive-ins and a riding stable." He says, of a difficult portage around a high dam, that it requires "a 150-foot, 45-degree ascent of the breast, which is made of loose riprap liberally mulched with broken beer bottles." He mentions a river town called Clearfield, "the only place I know of where you can pull up a canoe in an A & P parking lot." He describes a quill-rattling porcupine approaching their camp at night as sounding like "a set of asthmatic venetian blinds" and adds that, head on, "porcupines often look like reform candidates for mayor of Pittsburgh."

But along the West Branch he also notices the metamorphosis of the back country, which after having once been busy with prosperous mines and lumber camps and mill towns is slowly being reclaimed by nature. He describes "corduroy logging roads now rotting and being converted into compost; old mine shafts being filled and broken by slides; breached millraces that are now used by trout; collapsed bridges traversable only by raccoons; villages that were once prosperous towns."

Beyond the disintegration of commerce, he notes brilliant cameos of the wilderness: "A red fox vixen at the mouth of a den between rhododendron roots, staring intently as we drift under her bank without moving our paddles, or even our eyes. . . . A clump of azalea bearing perhaps the most brilliant flowers that ever bloomed, so bright that at a distance, illuminated by the rays of the early sun burning through the river mist, the blooming bush is mistaken for a flame. . . . Two young great horned owls, their heads still fuzzy with infant down, unsteady and clumsy on wing. They are accompanied by an adult on what may be their maiden flight. It probably is not a pleasant one for them. They are picked on by a mob of cursing crows. As we keep pace on the water, the big hunters are harassed and driven half a mile downriver from hemlock to hemlock."

And there are refreshing counterpoints to the almost overwhelming beauty. I mentioned earlier something about talking to a stranger on a bus. Gilbert talks to strangers everywhere, on buses, trains, planes, elevators, restaurants, sidewalks, mountaintops, islands, mineshafts, everywhere. I won't be gushy and say that he loves people, but he certainly does find them interesting, perhaps because he is open and receptive and listens to them as they talk about themselves. His account of the beauty of the West Branch is accented somehow by an incongruous encounter one day on shore with a modern-day hippie, described by Gilbert as "a fashionably bearded, sandaled, bandannaed young man who was carrying a three-foot pine plank." "Wow, far out—a canoe," the young man said. "I'm into martial arts in Altoona but . . . I told the old lady I had to get out in the country and get my head together. I ripped off this board and I get anybody I see doing anything cool to sign it. You have to sign."

A hippie in the wilderness. As valid a part of the environment as the owl and the fox and the azalea bush; the slaughter of the buffalo and the deer; the disappearance of the ferret and the paddlefish; the prevalence of worms; the intense interest in vastly different living things professed by Jack Lynch and Ray Edwards.

They're all part of Bil Gilbert's world. Warning: you can get caught up in it. I don't see how you can avoid it. Or why you'd want to.

A final word or two about Bil. I will grasp the nettle and discuss the matter of the first name. Some people who don't know him worry about the single ell, feeling that it is pretentious or commercial, an artificial shortening to attract attention of one kind or another. Not at all. It's an acronym derived from the first initials of family names, and it's been Bil's since his parents gave it to him at his birth. His full given name (I don't believe he keeps it a secret) is Bil Leroy Gilbert.

As for pretension, on a scale of 1 to 10 Gilbert ranks somewhere around zero-minus. I never knew a man less concerned with what people think of him. I don't mean he defies convention and opinion; they just don't matter that much. Take hair, which occupies so much of our country's attention nowadays. Some people say that Gilbert wears an old-fashioned, flat-top crew cut. Others insist that he has an unruly brown mop that tumbles down over his ears and collar. Both sets of observers are correct, but they're like ornithologists seeing different plumages at different times of the year. Gilbert usually gets his hair cut every ten or twelve months. After the haircut, he's right in style with those avant-garde reactionaries who are trying to bring back the burr cut. A few months later he's more or less in step with straight-arrow America, his hair trim and reasonably neat, not too short, not too long. Some time after that he's in tune with the looser, longer-haired types. Then the stuff starts to get in his eyes and after clipping it away a few times he breaks down and heads for the barber again. His wife, Ann, a treasure (who often does the haircutting herself), and his four grown children, more treasures, think his hair is just fine, all the time.

His general appearance is that of an amiable blue-eyed bear. He's kind of round and full and tan, with massive arms and legs and a comfortable belly. In his Baffin Island story he says of himself and his large friend Walmer, "Sam and I both take some pride in liking cold weather and being resistant to it. Shivering acquaintances say it is because we are both too ample; we claim it is because we are sensible feeders who maintain proper substance and circulation, and that our critics would do the same if they were not so obsessed with being thinner than God intended our species to be."

His usual costume in summer consists of moccasins and low-slung khaki shorts, which give his stomach a chance to get some air. He owns a sport shirt, which he wears on more-or-less formal occasions, like going to racetracks. He has what he calls his Penney-Missouri Awards suit, which he wears to luncheons at which he is awarded trophies for stories he has written—a tan corduroy suit with which he wears a blue shirt and a bow tie. Otherwise, when he ventures into New York or Washington or Lincoln, Nebraska, to check up on what editors are doing to his stories, he wears high shoes, socks, trousers, a belt, a shirt that buttons, and a kind of greenish jacket. I believe he also owns a four-in-hand tie, although I'm not in a position to testify to that. He always looks comfortable when I see him (except at Penney-Missouri Awards luncheons), and I don't believe he'd be comfortable wearing a four-in-hand tie.

He works harder than any other person I've ever known, and he also has more fun. He's always doing *something*. He was born and raised in Michigan, where he lived on a golf course for part of his childhood (and developed into the best young golfer in the state; he's since given up the sport, probably because it's too static for him and takes up too much time). He played football in high school, later coached track and tennis and basketball, and spent his honeymoon with Ann on a bicycle tour. He lives part of the year in Pennsylvania, part of it in Arizona, and is almost always going somewhere else to do a story. How he finds time to write as much as he does as well as he does is beyond me. But I'm awfully glad he does.

Robert W. Creamer
Sports Illustrated

Part I: Other Bloods

It is a matter of fact that we have been—at least for as long as there are records having to do with human thought and speculation—insatiably curious about the nature and activities of other creatures. Should we ever come to a good or even more approximate understanding of why we are so fascinated by the affairs of other bloods, I think we will have much better answers than we do now about questions of what we are, how we got this way and what we might become. Whether having such knowledge would be good or bad for us is another matter, and one about which I am agnostic.

The Devil in Tasmania

The English natural philosopher and novelist C. S. Lewis wrote that we all yearn to know other bloods, the other creatures of the world, and that this is a singular and definitive characteristic of our species. Coming as I do from a family in which this yearning was encouraged, I was in complete agreement with Lewis long before I ever read him. I grew up in a household that always included other bloods—dogs, cats, mules, cows, mice, newts, raccoons, crows, bears, badgers—and I had parents and other adult kin who were hell-bent on pursuing a great variety of zoological interests, practical and theoretical, recreational and vocational. Observations, speculation and arguments about the ways of animals, birds, fish and so forth were as frequent and taken as seriously in our house as discussions of politics, sports or economics are in other families.

Some 45 years ago my father gave me a good reference book on the mammals of the world. Within a year or so I had practically committed it to memory, and while doing so, I became acquainted with the Tasmanian devil. Sitting in the boglands of southern Michigan, I fixed this odd, oddly named creature in my mind as the symbol for the zoologically exotic and mysterious. In much the same way, its native haunt, the island of Tasmania, came to represent for me the absolute in foreign geography. Yearning after the sight of this beast and its island did not, I think, become an obsession or the ultimate ambition of my life. Nonetheless, the notion persisted that, if things worked out right, someday I would cross the world to Tasmania, search for the devil and, if lucky, look upon one. Things did work out right, and I recently made my way to Tasmania, accompanied by my friend Sam Walmer.

Tasmania is the most southerly state of Australia and very nearly the southernmost outpost of the inhabited world. Because there's nothing much south of the island except penguins and students of ice, residents of Tasmania assume, reasonably enough, that anyone who finds his way there has specific reasons for doing so and hasn't stopped by casually on the way to someplace else. In consequence, Tasmanians, while polite, are persistently curious about the motives of travelers, particularly one such as I, who was a bit reluctant to admit he had come some 10,000 miles merely to look at an oddly named animal.

Jean Taranto is more or less professionally curious about visitors. She works variously as a wholesale travel agent, a publicist and a marketing specialist for tourist enterprises. A former BOAC stewardess, she settled in Hobart, the capital of Tasmania, because, she says, "There's no place nicer." Her forte, as she puts it, is "organizing things," parties, dinners, meetings between people who should know each other. "The American consulate uses me a bit for things of that nature," she says. It was because of this and an introduction from a mutual friend at the U.S. Embassy in Canberra, Australia's capital, that our paths crossed. In a seafood bar overlooking Hobart harbor she asked Sam and me what our business was and how she might advance it.

"Well, it's a kind of a free-form expedition," I replied. "Sam here is an apple grower in Pennsylvania, and he wants to do some research on your orchards."

The large, hairy young bloke in question was at that moment vigorously researching a sizable pile of Tasmanian crayfish. By temperament Sam is loud, iconoclastic, disrespectful, disputatious, observant and curious, all of which makes him a stimulating traveling companion. His occupation permits him occasionally to go off on eccentric quests to improbable places. Most important, because of his physique and a lot of rigorous on-the-job training, he's conditioned to pick up and lug very heavy things that his seniors find tedious and undignified to carry.

Tasmania is known as Apple Island because it used to produce most of Australia's apples and still grows about 25% of them. Therefore, though she hadn't previously met one, Jean didn't find a visiting pomologist implausible and didn't think organizing a tour of local orchards would be difficult.

"And yourself, Bil," she asked, getting on with it. "I'm told you're a journalist. Are you working on this trip?"

"Tell Ms. Taranto about it," Sam suggested maliciously, surfacing from a midden of crayfish parts. Because of my recent experiences with Australian consular, customs and immigration officials concerning "reason for visit," he was anticipating being entertained by the rest of the conversation.

"Well, quite often I write about natural history," I said, scrambling. "Tasmania is very interesting in that way — so many endemic species, parallel evolution and so forth. . . ."

Jean didn't get to know what nearly everybody in Tasmania is doing by being put off. "Of course," she said, "but what is it you'd like to see?"

There seemed no way to avoid the truth any longer. "There's a *lot* I'd like to see, the rain forest, wombats, platypuses," I said. "But I — or rather we, Sam and I — what we'd most like to meet is a Tasmanian devil."

"The little beast in the cartoons?" Jean whooped with surprise. "I have a friend in films you should meet."

I shook my head.

"I may be one of the few people in the English-speaking world who has never seen that cartoon," I said. "I never even knew it existed until my daughter told me a Tasmanian devil plays straight man to Bugs Bunny. No, it's the real animal we want to find, the one out in the bush."

Jean raised her eyebrows.

"It's really not all that weird an idea," I said. "Like kangaroos. I'll bet nearly every tourist who comes to Australia wants to see a kangaroo. You people advertise the hell out of them in your travel propaganda."

"We do, rather."

"My thing is about the same, only more specialized." I offered some disjointed biographical notes. Included was some digressive and gratuitous information on how and why to domesticate a badger. Also some confused mention of C. S. Lewis.

"Gorgeous," said Jean soothingly. "You explain it beautifully. Sometimes I'm so slow."

Jean Taranto isn't slow. Whether she had been persuaded of my interest in Tasmanian devils or had decided she could go along with a gag as well as anybody, she phoned the next day to say that she had an acquaintance who might be helpful, a young man named John Hamilton, a former journalist. A year or so before he and his wife had bought property on the coast, 60 miles south of Hobart. On part of it they had built a private zoo, to which they charged admission. Jean assumed that the Hamiltons had devils there, because the name of their enterprise was the Tasmanian Devil Park. I said it sounded like a good place to start. It would be well to take a look at the animals in confinement before we went thrashing around in the bush after feral ones.

The Hamiltons, we discovered, did have devils, a pair of yearlings that had been trapped by the Tasmanian wildlife service. During the daylight hours they crouched in the far corner of a fenced enclosure, glaring balefully at the cash customers. With a wry chauvinism, Tasmanians often claim that their devil is the ugliest animal in the world. Esthetic judgments are subjective, but it's understandable why this one is commonly held. At a distance—from which devils look their best—they are merely undistinguished, being low-slung, stumpy creatures covered with jet black hair sometimes splashed with white blazes across the chest and rump. In conformation, they somewhat resemble an ill-formed bear cub or wolverine. Closer examination destroys these and other analogies. A Tasmanian devil doesn't look much like any other single species but rather like bits and pieces of several stuck together without regard for beauty, symmetry or function. My own first flash impression when John Hamilton gingerly presented one for inspection was *mutant!*—of the sort that might proliferate in the aftermath of a nuclear war.

For the devil's size—a large one is three feet long and weighs between 20 and 25 pounds—its head is enormous and would seem to fit better on a wolf or an alligator. For reasons to be considered shortly, the devil has evolved so that it's little more than a huge set of jaws set on a modest body. These jaws are studded with teeth that are not only exceptionally large but also numerous; a devil in good working order has 44 choppers, sometimes 46. A dog has 42, a cat 30. It isn't difficult to study this dentation. Somewhat like the python, the devil is so hinged as to be able to open its mouth very wide, and it does this often, being habitually slack-jawed and gapish. Also, it's steady drooler.

The devil has prominent, almost hairless, batlike ears, small mean eyes, the long, coarse whiskers of a big rat and piggish, dripping nose. Its body is lumpish, overlaid in maturity with heavy layers of fat. Its legs are bandy, with the rear ones giving the impression of being disproportionately long, lending a jacked-up appearance. The devil doesn't look to be—and isn't—agile or graceful. Its ordinary pace is a shamble. When it needs to move more rapidly, it lurches. Its tail is about a foot long, fat at the base, very nearly bare and pointed at the end like that of a snapping turtle. Unlike the tails of most mammals, it isn't as much a flexible appendage as it is a fixed extension of the body, hardly more waggable than a nose or an ear.

As with people, some animals—English bulldogs come to mind—are able to overcome ugliness with charming personalities. The Tasmanian devil isn't among these. Those who know the devil best claim that its behavior is more repugnant than its looks. Technically, the devil is a carnivore, but it isn't equipped to be a frequent or effective predator, and certainly not a bold or dashing one.

We met up with only one person in Tasmania who had directly observed a devil committing a true act of predation. This was a park ranger, Oliver Vaughn, who was based at an isolated station in western Tasmania. One winter, when the snow cover had been deep, Vaughn's wife had begun feeding wildlife around the cabin. One morning a wallaby (the medium-sized kangaroo: about 3½ feet tall and weighing 42 pounds) was sitting upright in the snow munching on a slice of bread when a devil lurched out from under the cabin and grabbed the wallaby. "He seized it by the throat," Vaughn recalls, "appeared to kill it immediately and commenced to bolt it down head first. The width and power of those jaws is remarkable. Normally a devil wouldn't be able to grapple a wallaby, but this wallaby was perhaps made unwary by its hunger or handicapped by the snow."

Ordinarily, devils are scavengers and—to give them their due—extremely effective ones. Keen senses of hearing and smell enable them to locate edibles quickly, and almost anything they find they can grind up with their powerful jaws and teeth. Stockmen say that devils will completely consume a dead cow or sheep, eating bones, teeth, hooves, and horns. Such scavenging feats aren't performed by a single animal but by groups of a dozen or more, although they don't travel in packs but are drawn one by one to carcasses. They behave like a brawling mob, having, so far as anyone knows, virtually no social organization or restraining instincts.

While not attack animals, devils will take living creatures that are too young, old, enfeebled or immobile to escape them. In Tasmania there is a sizable commercial trade in the skins of wallabies and the silky furred native possums, and trappers have to get to their snares almost as soon as they are sprung to beat the devils to the catch. On one occasion, a sheep farmer on the northern part of the island brought his animals into a shearing shed with a slatted floor, underneath which, unbeknown to anyone, several devils were lurking. They weren't discovered and disposed of until several sheep whose feet had slipped through the slats had had their lower legs gnawed off.

Several times a day Hamilton puts on a "devil show" for the benefit of his customers. He enters the enclosure and gives a lecture on the devils' ferocious habits. Then, while making it clear that he's attempting something of some risk, he lures or teases one of the captives out of its corner and picks it up by its tail, which makes a convenient and safe handle, the animal being incapable of swiveling back to get at the hand that holds it. The devil thus held does the best it can, hissing, gasping, drooling and gnashing its teeth. Now and then it may give a wavering screech, a captive version of the wild, eerie call that gave the species its popular name. Before the first settlers were well acquainted with the screech's maker, the sound coming out of the bush struck them as being truly devilish. After seeing more of the animal, there seemed no reason to change the name. Hamilton's devil demonstration pleases his crowds, confirming their beliefs about the savagery of the beast.

There is another private zoo in the northern part of Tasmania operated by Peter Wright, a former professional diver, and his wife. The Wrights also have devils that have been presented to them by the wildlife service, but they raise theirs by hand, more or less as pets. Their devil show involves finger-feeding the animals and picking them up and cuddling them, an act the animals tolerate pacifically. The Wright's exhibition is also well received as a marvelous demonstration of handling a savage beast.

Devils aren't objects of general affection like the koala or the kangaroo, but Tasmanians are quite proud of them, partly because of the worldwide fame of the cartoon character. The feeling is that this is at least one thing the island is known for elsewhere. One result is that the devil is the best-publicized animal in Tasmania. Its head is incorporated in the insignia of the state park and wildlife service. T shirts bearing a caricature of a snarling devil (and the legend I AM A TASSIE DEVIL), devil drinking glasses, devil postcards and place mats, and even a Tasmanian devil board game—a version of Monopoly—are sold in gift shops.

Despite this exposure, many Tasmanians have never seen a living devil, except in the roadside zoos, and generally know as much about them as do people in Trenton, N.J. Some Tasmanians share an opinion held widely in the rest of the world—that their native scavenger is rare and perhaps on the verge of extinction. This is emphatically not the case. The Tasmanian wildlife service estimates that there may be a million devils on the island and, if anything, the population is growing too rapidly.

Tasmanians are as outdoorish as other people and at least as observant. The reason they seldom see and know so little about the devil is that the species is an extremely secretive one. It is thoroughly nocturnal, lying up in hollow trees or well-hidden dens during the day, and its coal-black coat, on which the white markings simulate shadow patterns, makes it all but invisible in the nighttime bush. Despite its ferocious appearance, it's a very shy animal, so shy that the aboriginal Tasmanians called it the Cowardly One. Its inclination is to hide when alarmed. Lumberjacks, trappers, hunters and game wardens say that while they often hear and smell the animal—along with everything else, devils stink because of glandular secretions and their line of work—they seldom see one.

The first person Sam and I met who had had much to do with feral devils was the owner of a sheep station in the Tasmanian midlands. His name was Digby, and eight generations of his family have lived and worked on the ranch. At present, Digby has 16,000 sheep and a good many devils, so many that the previous year the wildlife service permitted him to trap and destroy 50 of these normally protected animals. Even so, Digby and his sons say they may go a year or more between sightings of a free-roaming devil, and then it's usually a matter of catching one of the animals briefly in headlight beams while driving at night through their pastures.

Digby doesn't despise the devil with the passion that American sheepmen feel toward coyotes. "They're more of a nuisance," he says. "They'll take occasional domestic fowl from their roosts. A neighbor lost a litter of good shepherd puppies—the bitch left them for an hour or so and the devils got them. But while some believe otherwise, I've never seen evidence of them killing a fit ewe or lamb. They'll take a weakened or dead animal and make very short work of it. By morning there'll be nothing left but the fleece and the plastic ear tags. We have a notion that they may well be of minor benefit, because they dispose of diseased stock so thoroughly and quickly. We trap to keep the numbers down a bit, on the chance that if they did become too numerous and hungry they might turn to lambs. But we have no desire to eliminate them. They are part of this country. I wouldn't like to think there were none about."

Digby and his family live in a 150-year-old house built of native sandstone that has turned tawny with age. It's surrounded by a garden nearly as old, in which there are flocks of brilliant wild parrots and, unfortunately, a small grove of apple trees, which caught Sam's attention. After what I felt was an interminable discussion about the diseases of apples, Digby returned to the main subject and made an unexpected comment.

"The devils are rather unpleasant acting little beasts but perfectly harmless. I couldn't imagine them, for example, rushing out and savaging a man's leg. But there is another matter I sometimes think of. If I were to suffer an accident in the bush, go down and not be able to move, I'd be absolutely terrified of being found and taken by the devils."

Many of those who took an interest in our quest suggested that we talk to Robert Green, curator of zoology at the Queen Victoria Museum in Launceston, the second-largest city on the island, and author of an excellent guide to the mammals of Tasmania and of several monographs on the devil. Green showed no surprise at our purpose and, in fact, seemed to find it curious that he had met so few of our kind. "Marvelous, fascinating creatures," he said. "It's a pity we know so little about them."

Green, who's in his 50s, said that during his entire career he had observed wild, free-ranging devils on only five occasions. However, he has handled more of them than perhaps any other person, having set up zoological shop at sheep stations where they were being trapped. There he examined and performed autopsies on their carcasses. He's particularly instructive about basic demonology, what the devil is physiologically and how it got that way.

Like many endemic Australian species, the devil is a marsupial, one of the order of pouched, non-placental mammals. Except for the New World possums and a few of their minor associates, marsupials are now found almost exclusively in Australia and on adjacent islands. In ancient times they were plentiful in both Americas, Asia and Europe. Marsupials seem to have originated on one of those continents, not Australia. Zoological conjecture is that they migrated Down Under some 70 million years ago, very likely from South America, across archipelagos that have since disappeared. They found the land empty of other mammals—probably for geological reasons, few species preceded or followed the marsupials to Australia—and in undisturbed isolation they put on an impressive display of radiation and parallel evolution. That is, they spread out ecologically, evolving so as to fill a number of functional niches which elsewhere are occupied by many orders of mammals. There are marsupials that look somewhat like deer and antelope (the kangaroos and wallabies), squirrels (the sugar gliders), woodchucks (the wombats), bears (the koalas), anteaters (numbats), lemurs (the cus cus) and many small ones that would pass as mice, rats and shrews. In prehistoric times there were cowlike marsupials and a marsupial lion.

One family of marsupials, the Dasyuridae, became carnivorous. It includes tiger cats and quolls, which are ferret-feline types, and the devil, which is thought to be an early, primitive model. The devil's maternal apparatus is, for example, rudimentary. Like all marsupials, devils give birth to what in placental creatures would be undeveloped embryos. Newborn devils are about the size of honeybees, but they have the capacity to crawl forward through the fur of their mother, to find and affix themselves to teats and to remain there for several months, growing as a placental infant would in the womb.

After their long postnatal journey, little devils aren't rewarded with deep, secure pouches such as kangaroos and other marsupials provide for their young. The devil's pouch is little more than a circular fold of skin with a central opening, almost like a shelf. As the young—usually four to a litter—grow, they hang out over this shelf and, as the mother lurches about the woods, are dragged behind, bumping along the ground. Occasionally, small devils that have fallen or been jolted out of the pouch are found, but most displaced infants are probably eaten by other devils, perhaps even by their mother. If ever there was a creature whose lousy adult personality can be excused on the grounds of a traumatic childhood, it's the devil.

Devils were once distributed throughout the Australian mainland, and from there migrated to Tasmania over a land bridge that periodically connected the two islands, the last time some 10,000 years ago. Subsequently, the species disappeared from the big island, possibly because it couldn't compete with the dingo, a true dog that became feral after being brought to Australia by aboriginal man. However, the devil thrived on Tasmania, which the dingo never reached, and was there in the early 1800s to startle the first white settlers. By the 1920s devils had become scarce, although human harassment wasn't considered a major factor in their decline. Green and other zoologists believe a distemper-type disease decimated the population. The species recovered from this epidemic and is now more numerous than ever before.

Even so, Green wasn't optimistic about being able to turn up a wild devil on short notice. Citing his own experience, he reckoned that we were about as likely or unlikely to find a devil in one place as in another. This gave us an excuse to go wherever we chose on the island. Because we heard that Tasmania harbored a lot of other interesting things and because the pomologist-porter was getting soft and uppity from overindulgence in the hospitality of apple growers, we headed for the wilderness of the western highlands.

Though only about the size of West Virginia, Tasmania is a marvelously diverse place, topographically, meteorologically and biologically. It is ringed with hard, white, usually empty sand beaches and spectacular rocky headlands. The eastern half of the island is a pleasant place of rolling pastures, woodlots and pretty country villages. To the west is a complex of jagged mountains that rise 5,000 to 6,000 feet. These ranges are buffeted by strong, persistent winds, called the Roaring Forties because of the latitude, which boil up in Antarctic regions and sweep in from the Indian Ocean. Because of this mountain windbreak, the eastern, leeward half of Tasmania has a mild Mediterranean climate, sunny, frostless and dry. The higher elevations that intercept the Roaring Forties are very wet—with 100 inches of precipitation annually in some places—and frequently cold. Blizzards can occur even during the summer months of December, January and February.

The highlands are set with muskeg-like bogs, cold lakes, deep canyons, sizable caverns, swift rivers and one of the most extensive temperate rain forests on the planet. There are dense stands of eucalyptus, tree ferns, myrtle trees, leatherwood, dogwood and sassafras, the latter two species being related only in name to the North American ones.

Impressive as the upper stories of these forests are, the ground cover is of more concern—and agony—to anyone trying to cross the highlands. It's composed of dense, resilient masses of windfalls, mosses, ferns, bushes and vines. The popular names of some of these growths are very suggestive; tanglefoot, needle bush, grass tree, pencil pine, horizontal scrub.

On the first day out, having gained a high, bare, alpine ridge, we spotted a small tarn in the direction we were headed. It appeared to have a nice sandy beach and an open grove of pines, and it promised to be a good campsite. A narrow trail reached this pond by a circuitous two-mile route, but from where we were it was only half a mile or so directly down to the tarn and there seemed no reason not to cut cross-country. Only some heatherish-looking moorland and a few brushy ravines intervened.

Because of our peculiar interests and our penchant for poor planning, Sam and I have had considerable acquaintance with bad scrub—the laurel slicks of the southern Appalachians, the high manzanita and chaparral of the Mexican border, the Arctic barrens, the cockpit country of Jamaica—but we were unprepared for what grew in those Tasmanian ravines. Generally, it was of the consistency of chicken wire and barbwire uncoiled and piled up to a height of four or five feet. At first we tried forcing our way through, but this left Sam, leading the way, gasping and exhausted, which would have been acceptable as far as I was concerned except that after he had broken through a few feet of brush, it snapped back and was as solid as before. We tried walking on top of it, which worked for a few steps, but then something would give way and we would plunge down into vegetative crevices and caves. The other option was to stay below and crawl forward in tunnels made by wallabies and wombats, creatures much smaller than we.

It took almost two hours to negotiate the half mile shortcut. When we reached the tarn we were bloody, bowed and convinced that we didn't want to do any more of that sort of thing. Thereafter, we remained on the trails, grateful for the aborigines, lumberjacks, bushwalkers or whoever it was who had cut them.

There are no full-blooded aborigines left in Tasmania, the result of an evil matter that haunts this pleasant island. When the first white settlers arrived, some 5,000 aborigines were living there. They were a small, dark, primitive, innocent, pacific people, but the whites immediately began to hunt them, partly because they were a nuisance and partly, quite literally, for sport. Michael Howe, a 19th century bushranger–Robin Hood figure, said he liked "killing blackfellows better than smoking my pipe."

The last of the aborigines on Tasmania was a woman named Truganini, the daughter of a chief. At about 16 she was a great beauty and was betrothed to a young man named Paraweena. The couple and another native were traveling by canoe with two lumberjacks, Watkin Lowe and Paddy Newell, when Paraweena and the other man were overpowered by the lumberjacks. The two native men were thrown overboard, their hands were chopped off when they tried to climb back on board, and they were left to drown. Truganini was taken, often, by Lowe and Newell and subsequently passed among other whites. But she survived. As an old woman and sole remaining member of her race in Tasmania, she was kept as a curiosity in Hobart, where she died on May 8, 1876. As she was dying she begged of a physician, "Don't let them cut me up. Bury me behind the mountains." However, her skeleton, the bones strung together, rests today in a coffin-like box in the basement of a Hobart museum.

Tasmania was the only place where European settlers accomplished the Final Solution to their problem with native populations, although it was attempted in Africa, Asia and especially the Americas. Truganini isn't an endemic Tasmanian ghost. She haunts the bushes of the Western world.

Tracts of the Tasmanian bush have never been explored on foot, and inevitably there are many stories about unknown things—animal, vegetable and mineral—that may be hidden in there. The most persistent speculation concerns the Thylacine, whose scientific name (*Thylacinus cynocephalus*) has become the common one for the animal sometimes called the Tasmanian Tiger or Wolf. Whether it still exists is a matter of conjecture, both popular and zoological, somewhat like that having to do with the Sasquatch in Oregon: The difference is that whatever their current status, Thylacines indisputably once did exist and were fairly common in Tasmania—but only there, in historic times.

The Thylacine was (to arbitrarily use the more conservative tense) another marsupial carnivore, about the size of a small German shepherd and of generally doglike appearance. It had a stiff tail, like that of the devil, and a broadly striped back suggestive of the tiger. Thylacines were true predators. Their habit was to pick up the trail of a kangaroo or wallaby and stay on it until they exhausted the speedier animal. After the white settlers arrived, Thylacines became at least occasional sheep killers and several thousand of them were rubbed out by bounty hunters. By the 1920s they had become extremely rare and shortly thereafter invisible, if not extinct. The last incontestably living Thylacine died in the Hobart Zoo in the mid-1930s.

Since then, organized searches have been mounted, but nobody has found a living Thylacine or produced a carcass, partial remains or a photograph of one. Nevertheless, in almost every country pub in western Tasmania there's a bloke who, if he hasn't personally met a Thylacine, has a mate who has found Thylacine tracks or shot one of the critters and chopped it up for lobster bait. At the other end of the same bar there is invariably a bloke who laughs at this as pure grog talk and will say that if there were Thylacines, he would have found them and made his fortune by guiding parties of bloody environmentalists to them.

Among the wild stories, there are enough plausible ones to have convinced many wildlife authorities that the question of the Thylacine's existence is still open. Green is optimistic that they've survived. He believes that hunting eliminated them from open areas and drove the remaining Thylacines back into the inaccessible bush, where they were further reduced by a disease similar to that which ravaged the devils. He says he has recently been receiving what he believes to be valid reports of Thylacine signs and thinks the animal may be making a recovery, but a slower one than that of the devil, because the Thylacines were less numerous to begin with and more widely dispersed by human harassment.

Green also points out that devils often associated with Thylacines, serving as jackals to their tigers. If there are still Thylacines, they are probably still being followed by devils. This may account for the lack of Thylacine remains, the scavengers presumably being as able to munch up a defunct Thylacine as anything else. We had formed a good opinion of Green's opinions — as well as one of our own about the Tasmanian bush, to the effect that it could hide anything up to the size of a rhinoceros. We saw no harm in looking about alertly for Thylacines. None showed up, but the possibility that they might entertained us as it does most Tasmanian bushwalkers.

Aside from three native snakes, all venomous, a variety of sluggish but mildly annoying bush flies, the devil and perhaps the Thylacine, Tasmania may have the best-looking and best-behaving wildlife in the world. There can be few better ways to commence a day than to awaken at dawn and find oneself being politely scrutinized by a wallaby with a joey peering brightly out of her pouch.

Nearly all the Tasmanian beasts are nocturnal or crepuscular, which accounts for their general look of wide-eyed innocence, and it's all but impossible to trail them to the lairs in the deep bush where they sleep during the day. Also, it's pointless. All that's necessary is to set up camp early, get the evening's cooking out of the way, find a smooth peppermint tree for a back rest and wait for the nice beasts to drop by. They start doing so about dusk in a very unshy way.

Pademelons look like miniature kangaroos but may be windup toys from FAO Schwarz. Wombats are about the size and shape of furry medicine balls with facial expressions like those of elderly academics.

Ringtails are velvety-furred, lemur-faced, raccoon-sized possums that hang by their tails from trees and make chirpy, cooing sounds. Bandicoots are rabbity creatures with very long pointed, bewhiskered, inquisitive faces. When pursued they hop about as if on pogo sticks, and they probably retire to C. S. Lewis' Narnia during the day.

The quoll may be the prettiest, most winsome animal in creation. It's catsized, with a bright foxy face and a fluffy tail. Its coat is a golden tawny color sprinkled with large white polka dots. Usually in pairs, quolls whisk about a camp and frequently sit up like prairie dogs to make chittering inquiries about the leftover noodle situation. As a matter of taxonomical fact, quolls are marsupial carnivores like the devil. When not charming tourists, they are holy terrors in regard to small mammals and birds and have some feeding and maternal habits that might make even a devil blush. However, such is the appeal of a nice face and figure that quolls are universally thought to be admirable animals.

In comparison with the wildlife that exists in most other places, Tasmanian beasts are so bizarre and appealing that sitting by a campfire watching them hop, scoot, amble and swing in and out of the dark bush gives one an odd sense of being in a parade of characters out of Lewis Carroll or Dr. Seuss. The classic example of this is the platypus, the believe-it-or-not creature—body of an otter, bill of a duck, feet of a beaver, venom of a rattlesnake, lays eggs like a bird but suckles its young—that has become the universal grade-school metaphor for the mysterious ways of nature.

Like a good many residents from places other than Tasmania, we arrived convinced that platypuses must be exceedingly rare and that they would be kept in guarded sanctuaries available only to the better class of Nobel laureates. We were soon informed that the animals are common and, though protected, go about their business in an unsupervised way in many of the island's lakes and rivers. This didn't change our original opinion, because to us the phrase "common platypus" seemed a contradiction in terms. It's not logical to expect to come across such a weird animal as readily as one might a muskrat.

Yet one evening we were sitting on a pile of driftwood above a small lake listening to the currowongs—an attractive Tasmanian crow that has a nice evening song—and waiting for the bandicoots to arrive when, without ceremony or fanfare, first one platypus and then another surfaced in the water below us. They floated unconcernedly under our dangling feet, looking just as weird as advertised. When they departed, Sam said in a very flat, matter-of-fact way, "Do you know what just happened? Two guys from Adams County, Pennsylvania, sat under a gum tree by a billabong and two platypuses paddled past. I absolutely do not believe it."

It would seem that the sighting of two platypuses, to say nothing of seeing kangaroos, wallabies, pademelons, wombats, possums, bandicoots, quolls and currowongs, would be sufficient, but the wants of animalholics are insatiable, and there remained the matter of the devil. Almost daily while in the bush, we found scats and tracks of our quarry. Given their numbers and the cover available to them, we may well have walked close by dozens of snoozing devils. However, our closest known encounter with any of them came late one night when at least two devils commenced screaming in a dense thicket of Antarctic beech bushes, squabbling, for all we knew, over a Thylacine kill. In the devil's call there are elements of a bobcat's screech, a large dog being strangled and the cry of a man who has just smashed his thumb with a hammer. The blending is unique, and when the sound wells up out of the midnight bush, it is much more impressive than when prodded out of a young captive animal. The quality is such as to make it understandable why, when they heard it in a new, strange land at the far end of the earth, the first settlers thought something much spookier and more evil than a medium-sized scavenger was lurking in Tasmania.

On our way into the mountains we had stopped by a small lodge located in an isolated clearing on the Pencil Pine River, six or seven miles from the ranger station at Cradle Mountain National Park. We had been told that the operators of this place put out their kitchen and table scraps in a clearing and that in the evenings a good many animals, including an occasional devil, fed there. This proved to be the case, but the day we arrived we were too early for the garbage show and, worse, too late in the day for a meal. However, after some cajoling, a makeshift "tea" was provided. It was an ordinary meal, but it was followed by a truly outstanding dessert, a concoction of pastry, apricots and whipped cream. I carried on about its goodness until its maker, a talkative and talented woman named Fleur, came out of the kitchen. Fleur said she had cooked all over Australia and had taken the position at the country lodge while waiting for a gentleman friend to make arrangements for her to start her own restaurant in Launceston. She said she had had a hard but independent life and had never lived in a slum nor let any man treat her like an old rag. Despite everything and being 64 years old, many people mistook her for 45, because she dressed and groomed herself modern. She apologized for the leftover food at tea and promised that if we were ever in the area again she could do much better.

I said that I bet she could and that as a matter of fact we were planning on returning in a week or so. Fleur said she did a nice rabbit in wine with a pumpkin casserole on the side. I said that sounded like a winner, especially if followed by some apricot delight. Fleur said, "If I do sigh so meself, me plum slice is a bit nicer."

I said plum slice it should be, and that unless we became hopelessly entangled in the horizontal scrub, we would be back to try it and look for devils. We were able to do both. The rabbits and the pumpkins justified Fleur's confidence, and though I would have bet heavily against it, the plum slice indeed topped the apricot delight. Laying in some reserve rations of the sweet, we went outside and began poking around in the underbrush between the lodge and the Pencil Pine River. At dusk a kitchen helper dumped several large cans of scraps in the usual clearing, and soon there were 50 or so wallabies, pademelons, possums, quolls and tiger cats (a slightly larger, stouter and less attractive version of the quoll) feeding on them. A half-dozen lodge guests came out to watch these creatures, and there was considerable loud talk to the effect that if everybody found a safe place and was very quiet, a devil might show up and do vicious things to wallabies. However, the wait proved long and boring, and after half an hour or so, the crowd dispersed. Sam and I were left sitting on eucalyptus stumps with flashlights, watching the moonlit garbage pile. About 11 p.m. the feeding animals became noticeably uneasy and drew back, and a devil, looking at first like a heavy, blobby shadow, arrived.

Overtly, the animal wasn't much different from those we had seen in the roadside zoos. However, a feral creature, even if it's only munching away on the remains of fricasseed rabbit, is always more satisfying to observe than one whose activities are restricted by captivity. The difference is somewhat the same as the difference between watching Lou

Gehrig play first base and watching Gary Cooper play Lou Gehrig playing first base.

Standing knee-deep in the trash, the devil buried its nose in it and steadily crunched away, occasionally clacking its teeth and making grunting, choking sounds. After it had settled in, the other animals moved back, without showing much concern for the living garbage Disposall in their midst. There followed an interesting behavioral vignette. Why it occurred is inexplicable. Despite all man's science and curiosity, the inner life of other bloods is more mysterious to us than the workings of the solar system. We therefore can describe things analogously, in anthropomorphic terms. What appeared to occur at the garbage pile was as follows:

A stout possum moved alongside the devil and turned to stare at it directly and deliberately. It was as if a decent citizen had entered a rough bar purposely to outface a notorious bully. The devil raised its big head until the noses of the two animals were almost touching and stared back in a puzzled way, as if trying to remember or figure out a formula for dealing with impudent possums. After holding its ground for a minute or so, the possum was satisfied, or perhaps grew bored with the confrontation, and moved off several feet and recommenced feeding. The devil held the same position and continued to stare at the spot where the possum had stood. Then, rather suddenly, it seemed to come up with the answer to its problem, something to the effect of: "Ah yes, what I am is a ferocious Tassie devil. I act savagely."

Thereupon the devil emitted a fairly savage squall, chomped its teeth and wheeled in a staggering pivot toward the possum, which long before this awkward movement was completed, leaped up onto a post. The other browsers did the same, swarming into trees and bushes, whence they studied the devil, which stood shaking its head dumbly. The incident suggested that the devil had some predatory inclinations, of which the possums, wallabies and quolls were aware. But they also seemed to recognize that the physical and intellectual limitations of the beast made it not much more dangerous than a falling tree.

The devil began feeding again, but after about 15 minutes stopped and abruptly lurched back into the bush, from which, almost immediately, there issued thrashing and caterwauling sounds. When they subsided, a second, larger devil emerged. Perhaps because it had become habituated to humans by eating their garbage, or because it didn't recognize or care what we were, or for reasons we wouldn't recognize as reasons, it walked directly to Sam and me, sniffed our boots slowly and then stared dully at our upper parts.

This animal may have routed the smaller one we had seen, but there had obviously been other battles in which it had been a loser, or only a Pyrrhic victor. One flank was scored with a deep, partly healed, suppurating wound. It had lost an eye and was left with a socket of knotted weeping scar tissue, which twisted its face hideously. It wheezed. Its jaws hung open. Its muzzle was covered with mucous, and its odor was rank.

Nevertheless, this was an extremely satisfying animal, in part because it was a trophy representing the successful conclusion to a considerable quest. The best thing about it was that it was completely and convincingly another blood, known for a brief moment more intimately than we thought we would ever in our lives know one of its kind.

C. S. Lewis expressed theological observations ecologically and vice versa. He was of the opinion that we seek other bloods not out of curiosity about their what-hath-God-wrought peculiarities but because we have a desperate need to know and recognize them as our peers and are delighted and comforted by innocent association with other of God's creatures. Lewis thought that since the fall of Man we have been tempted and seduced by clever but fallacious arguments that we have been set above the rules and rhythms of nature and charged with dominating it. To the extent that we have accepted that proposition—that man is the unnatural animal—we have been made the loneliest of animals, confused about our origins and divorced from the company of our peers. However, our persistent yearning for other bloods is evidence that we haven't completely succumbed to hubris and that we continue to resist dangerous claims about our superiority. Lewis' conception is an elegant, comprehensive statement of the web-of-life, we-are-all-in-this-together thesis currently well thought of by pop ecologists.

By and by, the battered devil, finding us either unsavory or unfathomable, turned away and satisfied its blood by scavenging garbage. Shivering in the midnight cold, we watched until it had finished and departed, feeling, as questing beasts, fulfilled in our blood.

The Second Keeper

Once, in northern Canada, I camped near a den of wolves and spent most of the softly lit Arctic night enthralled by the sight of three adults of the species feeding and playing with four woolly pekingese-sized pups. In Alaska, I saw a great black wolf pull down a caribou on the ice of a frozen lake. Another time, in the central Canadian Arctic, six of us were paddling along a river a day or so after a wildfire had left the vegetation along the banks charred. Out of a jumble of rocks stepped a Mackenzie Valley dog wolf. He stood perhaps three feet at the shoulder, and his coat was snow-white, and he stood out against the blackened landscape like a monument. For a mile or so the animal trotted along the shore following us. Nothing in his manner suggested alarm or hostility. It was quite possible that he had never seen either men or canoes and, while it is an anthropomorphic judgment, he had about him an air of joyful curiosity.

I have had few other moments that I recall so clearly as those when I saw wolves in the wild. A good wolf is esthetically magnificent, even awesome. Like rare works of art or extraordinary athletic achievements, wolves tend to inspire transcendental thoughts.

Man's reaction to wolves is evident in the symbolism that surrounds them. The wolf is right up there with the lion, the tiger and the eagle as a metaphor for valor, strength and freedom. Wolf images adorn coins, seals, battle standards. We assign them honorable, or at least admirable, characteristics; compare being called a wolf to being described as a dog, a jackal, a coyote or even a fox. Though some wolves may go bad, like those who pestered the Little Pigs and Ms. Hood, we tend to feel about them as we did about John Wayne. They may be prickly, easily roused and dangerous, but they are essentially proud, true heroes. Even in fantasy it was unthinkable for Mowgli to have been adopted by jackals or the red dogs of the Decan; these lesser canines were sneaky villains against whom the jungle boy had to be vigilantly protected by the gray heroes of the Seonee wolf pack.

In practice, of course, we have treated wolves as if they were crosses between boll weevils and Martian invaders. In this country, clearing the place of wolves was a work that European settlers kept at for the better part of three centuries. For good or bad, real or imaginary reasons, we set after wolves with guns, traps, poisons and even biological warfare. Captive animals were infected with mange and released to mingle with wild wolves that, it was hoped, would become hairless and freeze to death.

The final solution to the wolf problem was not accomplished until the first quarter of this century, when federal and state agencies fought the last of the wolf wars on the Great Plains and in the Rocky Mountains. Now only a few hundred timber wolves survive in national park and forest sanctuaries in northern Michigan and Minnesota. There are occasional reports of wolves that, like defiant stragglers from a long-defeated army, may be hanging on somewhere in the northern Rockies. A red wolf or two may survive in the Southwest. Otherwise, the U.S., once inhabited from coast to coast by eleven subspecies of the animal, is wolfless, although there are still sizable wolf populations in Canada and Alaska.

There is an irony to our elimination of wolves. Once we had rubbed them out—even as we *were* rubbing them out—we began to miss them. Oldtimers who had taken part in the westerly movement began to note sorrowfully that the old, free days were gone; now there was only Omaha and all it implies. That a man could no longer hear a wolf howl was seized on as the sad symbol of these changes. Nostalgia for wolves grew and in our times has been reinforced by environmental considerations. As a result, even though the animals themselves are now of minimal ecological significance, there are an inordinate number of wolf students, of people trying to buy wolves and keep them as pets, of people looking for wolves to sell, of people deeply concerned with the survival of the few wild wolves left—in short, a lot of wolf buffs.

There is probably nobody more fiercely devoted to wolves than Jack Lynch, who has spent the last 19 years living with, caring for and guarding wolves. His good works have extended to several subspecies, but he is principally, almost religiously, concerned with the fate of the most renowned of all North American wolves, the almost mythic lobo, the buffalo wolf, *Canis lupus nubilus*. Lynch, 54, lives in a fenced compound on the Olympic Peninsula of the State of Washington. Along with representatives of five other wolf subspecies, he shares the place with 72 buffalo wolves. They are, it is generally thought, the last of their kind, there having been no reliable reports of *nubilus* in the wild for almost 50 years.

During the 19 years, Lynch has reduced himself to, or a bit below, the poverty level, has given up most of the comforts and necessities that are considered ordinary, has parted from his wife and has worked himself into a chronic state of exhaustion in order to do what he feels must be done for the wolves. Beyond the posthole digging, fencing, shoveling, cleaning, doctoring, question-answering, money-procuring and inspection-resisting (a particular nuisance), there is the matter of food. Lynch, who knows as much about wolf nutrition as anyone ever has, says a mature wolf needs 35 to 40 pounds of meat a week to remain fit. This cannot be chopped liver, so to speak, much less the dried, processed meal fed domestic dogs. Lynch becomes very agitated when he talks about what grain will do to wolves, which do not have, he says, the enzymes to handle it, and about people who try to maintain wolves on this sort of feed. According to Lynch, the meat must be real meat, with blood, bone and guts, of the sort buffalo wolves got for themselves in better days. Because during these 19 years he has seldom had the money to pay for a sufficiency of meat, Lynch has scrounged for it—around slaughterhouses, on ranches and farms where stock may go down, especially along roads for deer and other animals killed by cars and trucks. Upon discovering a carcass, often a high ripe one, Lynch will butcher it—covering himself with gore, offal and flies in the process—and lug the meat home to his wolves. Under such circumstances, supplying a single wolf with 35 to 40 pounds of meat a week for 19 years might be considered at least a semi-Herculean labor, but since he started keeping wolves, Lynch has never been responsible for fewer than 32 of them, and he currently lives with 98, of which 80 are mature. For nearly all of the 19 years he has done this by himself—a job that a century ago would have occupied 98 wolves. Year after year, come heaven, hell, high water, lawsuits, divorce, bill collectors, sprained backs and the flu, he has scrounged and dragged back the meat, a ton or so every bloody week.

There may have been only one other man able to fully understand what Lynch has done and why he has done it. That would have been a tiny, pugnacious physician named E. H. McCleery, now nearly two decades in his grave. He was the First Keeper, who for 40 years kept the ancestors of Lynch's lobos. More or less on his deathbed, McCleery, acting like an ancient pagan priest appointing a successor to guard sacred trappings, turned the buffalo wolves over to Lynch and charged him with protecting the gene pool, which both men regarded as precious. The coincidences, some of which now seem to him almost fated, that led to Lynch's becoming the Second Keeper are much on his mind. He is aware of his own mortality and he is preoccupied with the questions of who will be the Third Keeper and from where—and *if*—he will appear.

Trained as an engineer, McCleery, a Pennsylvanian, went West in the 1880s but soon decided that part of the country was more in need of doctors. He returned East, earned a medical degree and went West again, establishing a kind of circuit-riding practice in Wyoming. Sometime before World War I he had an experience that was to change his life directly, that of the then-unborn Jack Lynch indirectly, and collectively those of all buffalo wolves. At a jackpot rodeo McCleery observed a "wolf bait," which had been put on the program as an added attraction. A trapped, hobbled wolf was thrown into a ring where cowpunchers and their dogs set about worrying, roping and racking it to death. Though a frail man, only a few inches over five feet in height, McCleery apparently went berserk and tried to halt the event. The cowpunchers shoved him out of the ring and suggested that if he didn't like their sport he could leave, not only the rodeo but also the territory of Wyoming, that they didn't need a doctor bad enough to put up with one who was a wolf lover. McCleery said he didn't need barbarians and sadists, and he packed up and went back East, finally settling in Kane, a village in the ridge country of western Pennsylvania. There he set up a practice and continued to brood about the fate of the wolves, which were then being exterminated throughout the West. Using his influence with former university friends who had risen to high places in the government and also through direct correspondence with professional wolf hunters, McCleery began to purchase live-trapped animals and have them shipped to him in Pennsylvania. Between 1920 and 1931 he obtained 25 wolves. Having become a knowledgeable canine taxonomist, McCleery determined that most of his animals were true lobos, a subspecies that would become extinct in the wild by the late 1930s.

(Differences between subspecies of any animal are very slight. A lobo, *i.e., nubilus,* is very similar to a northern Rocky Mountain wolf, *irremotus*. Some of the latter still survive in the Canadian Rockies and occasionally appear south of the border. Because the former ranges of these two wolves overlapped, presumably they interbred and interchanged characteristics. Because taxonomists, who are scientific classifiers, are hairsplitters by profession and therefore disputatious, McCleery's claims that his wolves were of *nubilus* stock were inevitably challenged—and they continued to be after Lynch took over. The controversy became so hot that in 1974 Ronald H. Novak of the U.S. Fish and Wildlife Service, one of the country's most distinguished taxonomists, entered it in the role of referee. Novak ruled that the McCleery and Lynch wolves, which he called a "unique national resource," had indisputably originated in the Great Plains area, where *nubilus,* the buffalo wolf, was endemic. If, and to what extent, they had intergraded with *irremotus* was a question Novak sidestepped adroitly, writing, "Regardless of taxonomics nomenclature . . . the group of wolves owned by Mr. Lynch contains genes no longer available in the wild, of a kind of wildlife that had a major role in the ecology and history of the western United States." The uniqueness of his animals having been established, Lynch, like McCleery before him, remains firmly convinced that they should properly be called *nubilus,* and he fiercely guards their genes against mongrelization and their good name against the sniping of taxonomists.)

For a time McCleery kept the wolves near his office in Kane. However, under pressure from his neighbors, he eventually moved them to a farm outside the village and in time gave up his practice to work with and for the wolves. They literally ate up his savings, and to maintain the animals and himself, the elderly doctor opened the farm to tourists who paid to see the wolves. In 1960 McCleery was 93, impoverished, ill, feeble and gnawed with worry about what would become of the lobos when he was gone. At this moment, for reasons he still finds a little spooky and hard to explain rationally, Jack Lynch arrived in Kane.

Originally from Illinois, Lynch was a World War II veteran and subsequently a bush pilot, demolitionist and prospector in Mexico. In the 1950s he had settled in Milwaukee, working as a superintendent for a large heavy-construction firm. Lynch says the job paid well—and offered opportunities for making a lot more money. But, as had been the case throughout his life, he grew restless, dissatisfied, moody, sometimes violent-tempered. In search of stimulation, Lynch, who had always had a taste for macho sports with nostalgic overtones, began training and racing sled dogs. While flipping through a dog magazine, he came across a short article about Dr. McCleery and the lobo wolves he had in Pennsylvania. "I can't explain it," Lynch says, "but right then I had the feeling that this was something I had to see and know about. The next week I took a couple of days off from work and drove straight through from Milwaukee to Kane."

Lynch found the wolf farm a shambles. "Dr. McCleery was too old and weak to do any work or even supervise it," he says. "He had a local man taking care of the animals. I met that ———. He was carrying a club and pistol, and he told me, laughing, that those were his two most important tools. The pens were filthy and the animals were in bad shape, but I had never seen anything, no other animal or human, that moved me as those wolves did. Within five minutes I was absolutely sure that this was what I had been looking for, that taking care of the wolves would give my life some meaning."

Lynch returned to Milwaukee, and during the ensuing months he sold property he owned there, quit his job and, with his wife, returned in 1961 to Pennsylvania, where he purchased some of Dr. McCleery's farmland and all of his wolves—for $1,000 apiece. There were 32 of them.

"The first thing I did was fire that helper," Lynch says. "I didn't know anything about how wolves behaved or how to behave with them, but I knew there was something better than coming at them with a club or a gun."

Lynch began learning quickly. Three weeks after he bought the wolves he found that a 200-pound male named Saber had commenced to worry a hole in the fencing that enclosed him. "I knew absolutely nothing about wolves," Lynch says, "and I was too dumb to realize how ignorant I was. I'd had no trouble before this, and I'd half-decided that I was working with big dogs. I went into the pen and started fixing the fence without paying much attention to or, worse, without greeting Saber. The next thing I knew he had knocked me down with a shoulder block. [This is a common move when wolves are tussling in a semi-serious way with each other.] Then he put those huge jaws, which could crack a buffalo's spine, around my thigh, half-lifted me and began dragging me around the pen. I can't describe how scared I was, but at least I had sense enough not to try to fight. I kept telling him good things and complimenting him. If he had wanted, I'd have kissed him. He put me down in a far corner and drew his lips back in that kind of grin they have. I started to get up slowly. He knocked me down and half-threw me over his shoulder, like a wolf would carry a deer. [At the time Lynch weighed 220 pounds.] Just like animals, men will black out from fright. That is what I did, because I have no memory of the next few minutes. I woke up outside the gate. Saber must have released me. I probably crawled on my hands and knees, because they were covered with mud. Now the significant thing was that this had not been a true attack, because if it had been I'd have been dead. It was a test to see what I was, or what I would do, or something. When I got through shaking, I went up to the house and told Dr. McCleery. He didn't seem surprised; in fact, he seemed pleased. He said something I didn't really understand then. He said, 'Well, it seems as if I found the right man.'"

About four months after he took over the farm, Lynch was startled when all the wolves began howling at midday and kept doing so for 20 minutes or so in a way that he says he had never heard before and has not heard since. "I have no explanation," he says of the happening. "I just know what I heard. Suddenly I had the idea that Dr. McCleery was dead. I went up to his house and he was."

The extended family groups that comprise a free-roaming wolf pack are organized along rather rigid hierarchical lines. At the top are what behaviorists have come to call Alpha animals, either males or females that for reasons of strength, wisdom, experience or other undiagnosable traits of character are deferred to and followed by lesser wolves, the Betas, Gammas, Omegas, etc. One function of the system seems to be to suppress violent competition in such matters as feeding and the selection of mates and to promote cooperative action while hunting in groups and caring for pups. Alpha animals may assert their rights and discipline or instruct lesser animals in rough physical ways. However, it seems that because of their intangible sense of status or self-importance, they are normally inhibited from making serious killing attacks on their inferiors, somewhat as ancient samurai warriors would not deign to dirty their hands on peasants. Also, the lesser orders seldom challenge their betters, not so much, it seems, because of fear of direct reprisal but because such behavior would be a serious breach of the etiquette of the pack.

Though he had no formal training, Lynch had a strong intuitive sense that even among these Pennsylvania animals, so long removed from the wild, an inherited social order prevailed. His response was to try to understand the ways of the wolves rather than to reeducate them. The process of learning the meaning of various communicative gestures and vocalizations, discovering what constituted good and bad manners, understanding acceptable and unacceptable forms of discipline was a slow, empiric one. What he finally became to these wolves was a super-Alpha male, a position that perhaps no man has ever attained before, perhaps not even McCleery. One morning, I accompanied Lynch on his rounds among the wolves, and the nature of his status, the importance of it, was very apparent.

On the Olympic Peninsula, the wolves, descendants of the animals Lynch first met in Pennsylvania, are enclosed in a compound of about 10 heavily wooded and shaded acres. This area contains 30 separate pens that are separated by walkways. The pens hold varying numbers of wolves, from pairs to groups of six or eight animals. Penmates are selected by Lynch to create packs, or the beginnings of packs, in which desirable bloodlines are maintained. As we entered the main gate and began walking down the central alley, the animals rushed out of the enclosed thickets and dens to the fences. As a stranger I provoked some curiosity, but not much, the collective opinion obviously being that I was a retainer of no great importance. Lynch was the main man, the object of a lot of communicative gestures and very close attention. Nothing in the wolves' behavior gave the impression of hostile or cowed prisoners trying to get at or cringe before a warden. The mood, it seemed to me, was much more like that of a good-natured holiday crowd which has gathered to greet a respected celebrity, say a Pope or a President.

Here and there Lynch would go inside a pen to, so to speak, press flesh with his admirers. He stopped to scratch and remove a few handfuls of winter hair from a young male and to banter—it cannot be described otherwise—with the animal, to tease him about the sorry condition of his half-shed coat. An ancient female walked up slowly and stiff-legged, her hindquarters aching from arthritis. Lynch gave the matron a gentle massage and commiserated with her about her joints and the ills of old age. With Lummox, a 180-pound Alpha lobo, Lynch wrestled and rough-housed. The familiarity was permitted and appropriate, in part because Lummox himself is something of a baron. Such attention directed toward a less confident animal might have confused the animal or seemed threatening to it. "There is no such thing as *a* way with wolves," says Lynch. "They have individual personalities, and you deal with each one differently. What they think of themselves is as important as what they think of you."

At the end of the compound, in a very dense cedar grove, there was a pen occupied by a large, rangy male, Lupi, and a rather petite young female. She bounced exuberantly, looking, it seemed, for attention and some fondling from Lynch. Lupi, however, kept his bulk between her and the pen door and us. He was growling, deep in his throat, continually and menacingly. His neck mane was raised and occasionally his formidable canines were exposed in a snarl. "Lupi is exceptionally jealous," said Lynch, confirming the obvious. "Some pairs are not especially possessive, but he is like an old man with a beautiful, outgoing young wife. I'm trying to convince him I'm not a threat. That jealousy could make some real problems in working with them."

Then Lynch talked directly to the wolf: "Lupi, I don't like that. There is no need for that." He spoke firmly but with no trace of anger or impatience in his voice. In a soft aside to me he explained, "Wolves can accept discipline because among themselves they are social, disciplined animals, but they can also be humiliated, and when they are, they can be dangerous. I learned the hard way. Once, before I knew any better, I disciplined a male too harshly. The bad mistake was that I did it in front of his three sons. He was humiliated, and that was what he never really forgot or forgave." Lynch says he has received half a dozen or so serious bites, but always for good reason, because, out of ignorance or impatience, he had flouted important wolf taboos or social rules.

"Lupi," said Lynch continuing his conversation with the growling animal, "I want you to stand on that log so I can come in to see you." Slowly, so that the wolf could study each move, Lynch opened the heavy gate, continuing to talk to the wolf in a dignified tone. Lupi kept on growling, but somewhat less ferociously, and finally he stepped backward and sat down on a cedar log near the entranceway. He was tense and he quivered, perhaps from the strain of self-control, but he allowed Lynch to pat him several times respectfully and then to play a bit with the female, which seemed oblivious to the charged emotional atmosphere. After a few moments Lynch withdrew, continuing to speak soberly to Lupi as he did.

The process of becoming an Alpha leader, something of an honorary wolf, has been, Lynch says, the ongoing challenge of his work and also its principal reward. His primary frustration has been that as the sole support of the pack and its emissary to and protector against the outside world, he has never had as much time as he would like to observe, contemplate and be with the wolves. From the moment he assumed responsibility for them, the grinding labor has never ceased. In addition to the unending work of bringing back meat, Lynch was faced with a number of problems when he took over the pack that had reached crisis proportions because of McCleery's infirmities and the inadequacies of the hired help. The fences and shelters were in disrepair, and the pens filled with garbage and dung. Some of the animals were suffering from injuries and disease, in part because McCleery, though a physician, did not believe in much doctoring of wolves, on the grounds that the weak should perish and the strong survive. While agreeing that this is the natural way, Lynch felt that the surviving lobos were so far removed from the wild that supportive measures were necessary. He began to treat injured animals, to worm and inoculate all of them against distemper and other diseases. He was at first dependent on veterinarians but became impatient with their lack of knowledge, their methods—and their fees. Now he is largely his own vet and has come to be regarded as an authority, often being called on as a health consultant by others who are attempting to maintain wolves in captivity.

Many of the problems and all of the work were complicated by a chronic lack of money. Having spent nearly all his ready assets to purchase the property and the wolves, Lynch has been in debt almost from the start. Because he is a man of Spartan habits, the perpetual shortage of cash hasn't bothered him much as far as personal comforts are concerned, but it has annoyed him when the welfare of the wolves has been affected. "There was one time in Pennsylvania when everyone [that is, the wolves] needed a full gorge," he says. "I didn't have any meat, and I didn't know where to get any. Then, like an answer to a prayer, a farmer called and said that he had two cows down, that I could have them if I came and got them. The trouble was I didn't have enough gas in my truck to make the round trip, and I didn't have a cent to buy any. It was spring and there weren't many visitors. I waited around most of the morning and finally a couple came in and paid three dollars to see the wolves. I took their money, rushed them along past the pens, got in the truck, got some gas and went for the cows."

Because visitors were his only source of income—with wolf work occupying him 12 to 14 hours a day, moonlighting at another job is out of the question—Lynch had until recently continued the practice of charging visitors to see the animals, but he regarded this as an unhappy necessity and not a reason for keeping wolves. Feeling that the lobos are a priceless national treasure on the order of whooping cranes or the original of the Gettysburg Address, Lynch gives the impression that displaying them for money to casual gawkers is undignified and demeaning for both the wolves and himself. He also is not well suited temperamentally for public relations. He suffers fools and foolish questions—"How long do they live?" "Can they kill a bear?" "Do they have to eat meat?"—with, at best, obvious impatience. Petty sadists, people who poke and throw things at the wolves, even those who speak disparagingly of them, enrage Lynch. Many of his stories about such happenings end up with accounts of how he refunded the visitors' money and drove them out, sometimes physically, from the sanctuary.

Lynch also has a generally low opinion of casual wolf fanciers and keepers. He estimates there are now about a thousand wolves living in homes, loose or caged, as exotic pets; like the cults of tarantula or rattlesnake keeping, the practice seems to be growing. One consequence is that there is an increasing number of creatures that owners assume to be wolves, but are in fact wolf-dog hybrids. Lynch feels this is a terrible development that may, in time, taint or cast suspicion on the bloodlines of all captive stock. There are exceptions, Lynch concedes, people who have real wolves and real concern for them; but most wolf-pet experiments end badly for both man and beast, he says. "Most of these people are on ego trips to begin with," Lynch says. "The keepers get tired of it very quickly when they find out how much work and how little glamour there is in having a wolf—or even one of those hybrids."

His deepest animosity is directed toward agents who obtain wolves, legally or otherwise, and breed and sell them to private collectors, roadside exhibitors and TV and movie productions. "Rotten rip-off artists," is one of Lynch's printable descriptions of these traders. His negative views on "puppy mills" and the people who run them have made him, Lynch says, the object of abuse and even threats. He is also furious about a recent robbery of his files, in which a number of photographic negatives of the lobos were stolen and from which prints are now being made and sold.

Finally, Lynch has become embroiled in some essentially intellectual feuds with what he admits are serious but, he thinks, misguided wolf students. He has some impressive supporters among wolf zoologists, including Dr. Durward Allen of Purdue, whose research on timber wolves has made him perhaps the most distinguished American authority in the field. In general, though, Lynch tends to be less than enthusiastic about academics. He is critical of their lack of direct, personal experiences with wolves and resentful because he sometimes feels he is patronized because of his lack of formal training. He has had, he says, proposals from professionals who want to test and perform experiments with wolves that Lynch feels are inhumane and/or trivial. "So long as I am alive," he vows, "nothing will be done that is not in their best interests as individual living animals."

Lynch also believes that there are professionals who would like to take his wolves away from him, subvert his authority over the animals and plagiarize his life work. He says that one professor, whom he charges with wanting to become the "führer" of the academic wolf establishment, told him that he could raise a million dollars in foundation money for a new wolf sanctuary in which Lynch could be employed more or less as a head keeper. "Then he told me," says Lynch, "that if this worked out he would send a couple of graduate students to debrief me, which is another way of saying, stealing what I have learned. I laughed in his face. I'll never work with him."

When he is discussing wolves, Lynch shows certain characteristics often associated with them. He can be prickly, defensive and, as he himself tells it, sometimes violently hostile. Until he is convinced to the contrary, he tends to be suspicious of human motives and activities. He says that his greatest personal ambition is to find someplace where he can live with and enjoy the pack and deal with people on his terms rather than theirs. In fairness, it should be added that Lynch also can be a warm, witty, extremely hospitable man and certainly an enormously informative and entertaining one.

Nine years ago, in search of a more perfect sanctuary, Lynch moved his wolves across the continent to Washington. The old pens that had been in use since McCleery's time had become disease-infested to the point that sterilization was impossible. The long, cold winters in Pennsylvania were proving hard on the animals, and the reproductive rate was very low. Lynch made several flying trips—the first of them also marked the first time he had left the farm since coming to it a decade earlier—to investigate new sites and finally selected one on the Olympic Peninsula overlooking Discovery Bay, near the village of Gardiner. It is a forested tract, with a mild climate, close to ranches and farms where Lynch thought he might obtain meat. To raise the down payment, he sold some property he had inherited in the Midwest, as well as the land in Pennsylvania. He made the move in the winter of 1971 when the pack numbered 56 animals. One way or another all of them were tranquilized, something Lynch does only in the most critical situations. Twenty wolves were flown to Washington, with an assist from a friendly vet who listed them as German shepherds, the air-freight rates for dogs being considerably less than those for wolves. The other 36 were crated and trucked across country by Lynch and a friend in a single, difficult nonstop winter drive.

All of the wolves survived the trip, and as far as the physical condition of the pack goes, the transcontinental move has proved to be a good one. The new pens Lynch built are free-form, winding through the big trees, and are not neat in the symmetrical, institutional fashion of those in conventional zoos. They are probably more comfortable and commodious for the wolves and more effective at containing them. Wolves are not easy animals to pen, being strong diggers and leapers capable, as Lynch has learned, of clearing nine-foot fences when well motivated. The enclosures are exceptionally clean looking and smell better, as I told Lynch one day, that any wild wolf dens I've investigated. The animals are fit and vigorous and show very little of the pacing and aberrant displacement activity that often afflicts large caged predators. The reproductive rate has risen, the pack having almost doubled in size since the move from Pennsylvania. There is an old saying that a good way to learn the truth about a man is to look at his dog. If so, the condition of these wolves strongly suggests that they have lived under the care of a devoted, thoughtful and very energetic man.

The adequacy of Lynch's arrangements for the lobos has been certified by innumerable federal, state, local and private humane agents, whom Lynch has had about his place like some kennels have fleas. (Incidentally, Lynch has found that with proper diet, wolves rarely have ticks and fleas, and there is thus no need to use poisons to prevent such parasites.)

Given the unique circumstances of the wolf sanctuary, Lynch's notion that he has been a bit over-investigated is probably not unreasonable. In a great range of endeavors—providing social services, preserving natural resources—the general feeling seems to be that only public agencies or large private institutions are competent and trustworthy. It is somehow suspicious and unseemly for a private, unmoneyed individual to be the protector of the last buffalo wolves or, for that matter, of anything; if the things are worth keeping, some official body like the Department of the Interior, the Ford Foundation or the Sierra Club should be in charge. Certainly, this is sometimes the case, but not always. As one who has had considerable interest in comparing various arrangements for preserving rare, wild creatures, I can think of no public agency or corporate-style private institution that could have done a better job of saving the last of the buffalo wolves than McCleery and Lynch have. There is another point worth considering. Every buffalo wolf in the U.S. was once under the care or control of state and federal agencies. The only buffalo wolves still alive are the descendants of those animals that escaped the attentions of public officials and came under the private protection of McCleery and Lynch.

On the Olympic Peninsula, as in Pennsylvania, maintaining the pack has been essentially a one-man operation. Occasionally students or unattached young wolf lovers have turned up offering their assistance. Some worked out better and remained longer than others, but none stayed for very long. Some, according to Lynch, seemed to be in search of therapeutic effect that might be derived from grooving on wolves. Most did not have the stomach or strength for the work involved. There were limits to how long even the best of them were able or willing to continue. Then three years ago a woman named Mary Wheeler turned up, and Lynch feels that her coincidental coming was one of the best things that ever happened to him or the wolves. An active, outspoken, confident woman, a kind of Alpha female, Wheeler had raised a family in the Los Angeles area, had gone through a divorce and had been wandering around the West in a van with three goats and some camping gear, looking for a retreat and meaning, just as Lynch once had. In 1976 she bought a cottage overlooking Discovery Bay, a mile or so from Lynch's property. Though heretofore a city or suburban person, Wheeler was a passionate animal lover. "I was the one everybody in the neighborhood brought animals to, and usually left them with," she says. "Abandoned kittens, fledgling birds, sick dogs." Inevitably, after settling in on the shores of Discovery Bay, she was attracted to the wolf sanctuary and began hanging around it as a volunteer worker. Her initial reception was not cordial. Lynch was preoccupied with his usual wolf work and with what he describes as the bitter breakup of his marriage. "I was worn down, half-crazy and ready to flip out," he says of this bad time.

"He was like a man withdrawing from the world," says Wheeler, "or maybe someone turning into stone. Once in a while I got a glimpse of what he really was—a very smart, kind man—but not very often. I started to avoid the place because there was trouble there that I couldn't do anything about."

In 1978 Wheeler heard a conversation in the village store to the effect that the wolf man was dying in his house, unattended. "I said, 'My God, why hasn't somebody done something?' " she says. "I can't understand how people can just stand around talking about something like that."

She went at once to Lynch's cabin and found him in a coma, brought about, it was later diagnosed, by a combination of exhaustion, bad food and little of it, a viral infection and assorted anxieties. Wheeler took charge, nursing and feeding, cleaning up the premises and doing what wolf chores she could. When Lynch recovered she stayed on, working regularly with him and for the pack. Her special area of expertise has become the pups, 13 of which she hand-raised this summer. Often in the past, small pups would squirm through fences into adjacent pens or press too closely against such fences. Because prohibitions against violence among wolves apply principally within packs and not to those wolves over there, many of the young animals that Lynch could not foster by himself were killed by adults as trespassers.

Despite—in fact, partly because of—their success with new litters, neither Lynch nor Wheeler, who has become in effect an associate pack leader, is optimistic about the future of the sanctuary on the Olympic Peninsula. This year, for the first time, Lynch culled nine animals that were weak or defective. As he somberly admits, there are too many wolves. Destroying the animals under any circumstances is emotional torture for Lynch, and in the future he plans to let natural attrition reduce the pack. By not allowing them to breed, he hopes to eliminate subspecies other than *nubilus,* retaining only the true buffalo wolves. He believes somewhere between 35 and 50 of these animals will provide sufficient genetic variety to maintain good breeding stock.

Lynch sees other crises closing in around the present sanctuary. Farm and ranchlands on the Olympic Peninsula are being sold off to resort and residential developers, and he must travel farther and work longer to find meat. His chronic money problems have worsened because of his impending divorce settlement and because he has closed the sanctuary to fee-paying tourists. He took this step because the larger number of wolves has made it impossible for him to care for them and also supervise stray visitors. Predictably, Lynch does not seem to greatly regret the absence of the tourists. The sanctuary is now largely dependent on the largesse of 100 or so "wolf adopters," individuals and families throughout the country who make monthly donations for the upkeep of an individual wolf or the pack in general. To facilitate such donations, the sanctuary has been formally established as the nonprofit, tax-free Dr. E. H. McCleery Lobo Wolf Foundation. Even so, expenses remain greater than income. The wolves get by because of the willingness of Lynch and Wheeler to live on very little and in debt.

Lynch feels that he should find a better, more secluded retreat. And there is a dream that is never long out of his mind. In it the buffalo wolves are returned to part of their ancient range and allowed to live there as feral, free animals. "Technically, it is feasible," says Lynch. "I would need a full section of land in an isolated area. I would fence it into three large pens, say of 200 acres each. I'd put a good pair in each enclosure and start to provide them with live game. I would begin to harass them, shoot at them with pellets that sting but do not wound, set padded traps that hurt but do not maim, chase them on horseback, in Jeeps and on snowmobiles, teach them that man always means trouble, to be afraid of the sound, sight and smell of him. By the time their pups were mature they would be wary enough to be released."

The question that brings Lynch quickly down to earth, as it does others who contemplate the reintroduction to the wild of wolves and other large species, is where this release might be made. There has been talk that wolves might be brought back to Yellowstone or Glacier parks, both of which may now harbor a few animals as either casual visitors from Canada or as permanent, secretive residents. However, there is a doubt that reintroduced wolves would remain in a park and thus avoid confrontation with private landowners and domestic stock animals. Or, if they did, that their presence would be compatible with other park uses. Some ecologists have suggested establishing a large preserve in the northern plains and mountain area that would encompass 20 or 30 thousand acres of fenced land. In it wolves and other endangered species would be released and contained. The public would be excluded from these lands for several decades while the experiment was proceeding and its results analyzed.

"Biologically it is possible, but the politics make it improbable," says Lynch. "I doubt that this country cares enough or is farsighted enough for something like that. I don't really believe that the Parks or Forest Service or any government outfit cares enough to organize a non-public-use preserve or has guts enough to keep the long-range promises that would have to be made."

Because of his feelings about public wildlife agencies and agents, Lynch says flatly, "I would not permit any of the buffalo wolves to be used in any reintroduction scheme unless I had absolute, no-strings-attached authority over how they were used. Too much has gone into preserving them to take a chance on them being lost or mongrelized by some bureaucrat who cares more about politics and his job than about wolves."

Lynch has a backup ambition. "I'd like a place where, for whatever we have left of our lives, Mary and I could work quietly with the wolves and preserve the *nubilus* gene pool, so it would be there if in the future something happened that would permit successful reintroduction. It would be a place without near neighbors and the hassles that go with neighbors. God knows I'm used to living without money, but I'd like enough so we were not always on the verge of disaster—enough for some decent equipment and maybe to pay a living wage for one or two competent assistants. I'd like a lot more time just to observe and think. I've been with these wolves 19 years, and I'm just beginning to understand how much there is still to know. I've never had time to organize and expand my notes, or even make notes on a lot of observations and ideas I've had. On a place like that I'd want a small guest cabin where serious students, not the ego trippers, could stay and watch the wolves. We could exchange ideas. I don't know how or if that can happen. I don't think it should be done on public land or with public money. I suppose I'm thinking of a private patron, an angel or a consortium of angels. I keep hoping that something will turn up. The strange thing is that from the beginning, when old Dr. McCleery got the first wolves, something has always turned up when the wolves needed it the most."

As has been suggested, Jack Lynch is not a man who reacts agreeably if he thinks he is being baited or teased. The final question was put carefully. "Jack, you have given 19 years—hard years—of your life to these animals. As a matter of cold fact, they are museum pieces, like a lock of George Washington's hair or an old musket. They can't survive except as wards. They have no natural ecological function or impact. It seems very unlikely that anything is going to change for them. So the hard question is, why have you done what you've done? Why do you keep on being the Keeper?"

Lynch took a quick step or two backward, pivoted on his heel, stared for a few moments toward the wolf pens and then turned again, shaking his head from side to side. "What you are really asking is whether my life has been worth anything?"

"Yes."

"All right. If the animals in those pens disappear, the buffalo wolf is gone forever. There is nothing we can do to recreate them. In their genes is the history of the evolutionary and ecological forces that made this land—made us, too. I have kept something unique, historic and magnificent alive in this world. Otherwise it would be gone forever. Before I came to the wolves I was like a lot of people. I couldn't name one thing I had done that was worth anything to anybody but myself, that would have any meaning after I was gone. Now I have saved the genes of the buffalo wolves. I am part of history, which means I am part of the future, whether anybody remembers my name or not. I believe that my life has been worth something. I don't envy any man."

Addendum

By attrition, the Lynch pack has been reduced to about 60 members, all of whom it is thought are of the "buffalo wolf stock." In 1980 Lynch and Mary Wheeler moved with the wolves to an isolated 100-acre tract above the Yellowstone River in Montana. The present sanctuary is within the range where the ancestors of the present animals once lived freely. For this and other reasons Lynch and Wheeler think the present location is more suitable for the wolves and themselves. We stay in touch but, intentionally, I think of them only occasionally—for somewhat the same reasons I avoid excessive star gazing. Like pinpoints of light in the night sky, their work and passion hints at possibilities within the firmament of ourselves that I, at least, am not ready yet even to try to name, much less consider too long or directly.

The Missouri Kid

The range of most species coincides generally with the range of graduate students. So goes an aphorism used by wiseacre natural historians to point out that our knowledge of other creatures is neither so complete nor accurate as it is often presented as being. Neat, firmly drawn distribution maps in zoological texts will indicate that armadillos, mountain lions or pine voles inhabit a specific region because, within that region, observers whose credentials authorities accept have seen the beasts in question and perhaps even collected their hides and heads. In all probability, other members of the species are located in adjacent, similar habitat areas, waiting for the graduate students to catch up with them, but this is no more than an educated guess. An even shakier guess, however, is that the creatures are not in places outside the accepted range (nor in places not indicated on range maps).

When it comes to more complicated questions—why certain animals are where they are and not elsewhere; what their collective and individual motivations are and, if they have such things, their perceptions, pleasures, hopes, fears and ambitions—we are much further at sea. For working purposes, we proceed as if our reasonable assumptions and plausible explanations are, in fact, facts. Often we get away with this because we seem to have made shrewd guesses. But now and then something occurs that is so unreasonable and so implausible that we are forced to acknowledge another area of great mystery. Take some recent developments among the moose.

Mammals of North America, by Victor H. Cahalane, a distinguished zoologist and former chief naturalist of the U.S. Park Service, is a popular, well-regarded text. In it, the range of the moose is described as "the coniferous forests of northern North America; south of the limit of trees, from Nova Scotia and the Adirondack Mountains west to northern Minnesota, central Saskatchewan, southern James Bay, and the Mackenzie River delta to Bristol Bay and Kenai Peninsula of Alaska. South in the Rocky Mountains to central Wyoming, Idaho, and (occasionally) northern Washington."

Within this range, Cahalane writes, the moose is typically found in timber or wetlands foraging on aquatic plants and high-standing bush, "and usually spends its entire life in a relatively small area. The only time that a bull takes a trip is during the mating season, when he may bestir himself out of his little territory of five square miles to track down a cow or two."

More academic and technical works provide greater detail, but generally this is the official state-of-the-science moose line. It's probably a good enough one for most moose—but not for all of them. Events of the past several years demonstrate that it inadequately describes the potential of this species. In the Midwest the range of the moose must now be extended from the Minnesota-Ontario border—long regarded as the most southerly moose habitat—to the environs of metropolitan St. Louis. While, as Cahalane says, the average moose-in-the-woods may travel only a few miles from its home thickets, the cruising range of an individual can be 1,000 miles or more. Guidebook moose may continue to be content crashing around in coniferous forests, wading in bogs and feeding on pond lilies and sapling tops, but other moose, we now know, can maintain themselves in good condition and spirit on the relatively treeless prairies, can gracefully leap six-foot fences and can be happy foraging on multiflora roses, coralberries and winter wheat.

Moose lore, from now on, will be incomplete without this information (and a good many other curious addenda) because of the activities of an extraordinary animal, a Marco Polo of moosedom, a Magellan of its kind, who for more than two years has been wandering about the Midwest, puzzling zoologists and vastly entertaining the citizenry in a number of heretofore mooseless regions. His travels have been so remarkable and his adventures so picaresque that it simply will not do to speak of him here simply as a moose. He must be distinguished, as he has distinguished himself, from all others. Call him the Missouri Kid.

Like those of English rock groups, flying-saucer persons, Democratic presidential candidates and many other celebrious creatures who descend on us unexpectedly, the origins of the Missouri Kid are obscure, and it is now probably too late to do anything but speculate about what they truly are. A good many off-the-wall suggestions—that he started out from a commercial moose works or a zoo, or was a tame roadside attraction—have checked out negative. The only logical possibility remaining is that the Kid was conceived and born someplace in the Big Woods that stretch between Lake Superior and Lake-of-the-Woods along both sides of the U.S.-Canada boundary. If that is so, there is reason to assume—but no incontestable proof—that he came into the world in the spring of 1975.

"Calves are generally born in May and stay with the cows through their entire first year," says Pat Karns, a research biologist and moose specialist employed by the Minnesota Department of Natural Resources. (Minnesota has been the only midland state that needed a professional moose person, but there may be future openings in this field in other states. In the meantime Karns' zoological colleagues from states south of Minnesota have been calling him for background reports on mooseology.) "The next May, when the cow calves again, the yearling will leave," Karns says. "If it is a male, it quite often forms an attachment with a mature bull. They'll travel and forage together throughout the summer. In the fall, the old bull will enter rut, and he'll turn on the young male who has been tagging along, drub him and drive him away. That is when these youngsters often start wandering. Now and then we'll find them two or three hundred miles south of where they belong, in the farming country below the Twin Cities. However," he adds, referring specifically to the Missouri Kid, "the distance this one has traveled, assuming he came from northern Minnesota—and where else could he come from?—is unique, so far as I know."

Nobody saw what happened to the Missouri Kid in the early fall of 1976, when, presumably, an older male gave him the boot. Even if he had been under close surveillance, no one would have been able to enter his inner world and determine what motivated him. (Pat Karns' theory is simply a reasonable possibility—not a conclusion, but a logical point of departure from known fact. It is well to keep in mind the dictum about graduate students and all it implies.) Perhaps the old mentor bull was an especially scary one, or the Kid was a particularly sensitive adolescent. Whatever, that fall (maybe) the Kid apparently started legging it southward through central Minnesota. When and where he went is very guessy; this is the only portion of his subsequent expedition that cannot be documented with precision. Perhaps nobody saw him. If someone did, he was unaware that he had met a future celebrity.

It is a fact that a bull moose sporting an impressive rack of antlers showed up in mid-December of 1976 in north-central Iowa. He moved into some woody bottomlands along the upper Des Moines River near the village of Emmetsburg, a mile or so west of Five Island Lake. He settled down there temporarily, to munch river willows and delight local residents. Many came out to look at the odd beast, and some brief, friendly drives were organized to stir him far enough out of the thickets to facilitate photography and up-close admiration. Perhaps because of cartoon exposure, a bull moose, even though it is the largest and most powerful member of the deer family, strikes most people as a harmless and somehow humorous animal rather than the formidable one it can be. Throughout the Missouri Kid's odyssey he has been universally accepted as a charmer rather than a potential danger.

By the time he reached Emmetsburg, the Missouri Kid had made an exceptional trip for moose, but not yet a unique one. "Every so often, one comes down this far," says Lee Gladfelter, an Iowa deer biologist who was to become as much of a moose biologist as that state has. "Most of them hang around for a short time and then disappear. Maybe they work their way back north or get killed on the road."

According to Karns, the latter is often what happens to the few young bulls who wander down into the southern regions of their home state. "They meet with accidents," he says. "So far as we know, none of them has ever succeeded in getting back to traditional moose country. That doesn't mean one or more hasn't, or won't."

The last moose known to have visited Iowa, about five years ago, came to a very bad end. He was poached. "Some gunner just couldn't stand to see something that big and strange go on living," says Gladfelter. "He knew what he was shooting. We gave him a stiff fine, but that didn't help the moose. The one around Emmetsburg didn't have any problems of that sort. He became a local attraction and everybody became very protective about him."

The Missouri Kid remained in the willows near Emmetsburg for about nine months, until the early fall of 1977, when he suddenly abandoned this informal sanctuary and the host of friends he had made while there. "We began to get reports that he was moving down the Des Moines River, quite rapidly in a southeasterly direction," Gladfelter says. "It was as if he had a destination in mind, but who knows if he did, or what it was, or why he started traveling."

And that is the mystery—why? Perhaps the Missouri Kid was harassed by dogs. Maybe the flies became intolerable. Maybe he was driven on by the hormonal tides that wash over bull moose at that time of year. Maybe he started off to find a cow, but rather than going five miles or so, as proper animals of his kind are supposed to do, he kept on searching for the object of his desire for 500 miles.

"Nothing can be proved," says Karns. "We've had some terrible winters up here lately. Maybe he just had enough and decided he was going south."

It should be made clear that Karns is joking. Zoologists and other authorities will make such pleasantries to spice up a conversation, but they would be embarrassed if they were taken seriously—for good reasons. There is absolutely no evidence that a moose can analyze weather, remember one winter from the next and make long-range plans based on this information. But the reverse is also true: there is no evidence a moose cannot do such things. This is why the behavior of other species remains more obscure than the surface of the moon.

"Some animals of the same kind are smarter than others, or dumber," Karns jokes again and, in doing so, touches on another elemental matter—although very lightly, because professional prudence and tradition discourage such a concept being taken with the seriousness it logically deserves.

Academics, researchers, field people, naturalists, herdsmen, falconers, pet keepers, everyone who has had much intimate experience with other creatures knows beyond doubt that individual members of species have what we call, in ourselves, distinct personalities. They must, because the activity of a single creature reflects a combination of inherited behavioral responses and learning—that is, knowledge acquired through individual experience. No two have precisely the same experience (or, presumably, the same inherited capacity for acquiring experience and acting on it), and therefore, each personality is in some respects unique, whether it be in men or moose. However, to prepare even a sketchy biography of a single moose, to investigate the myriad formative influences, would be far more difficult and time-consuming than chronicling the life of a defrocked politician. Also, it would be practically impossible, communication alone being (for now, anyway) an almost insurmountable barrier between man and moose. Finally, there is little professional incentive to make such differentiation. In practice, therefore, we generally treat each species other than Homo sapiens collectively, noting common characteristics and proceeding as if the behavior and potential of each individual is identical. We know this is not the case, but do not know how to work the personality factor into our studies, all of which are distorted by this omission. The distortion is implicit in our judgments and predictions and generalization about other species, which are upset when an individual like the Missouri Kid comes along.

Inexplicably driven, the Kid came loping down the Des Moines River in the fall of 1977. By late October he had reached the vicinity of Boone, about 150 miles from that year's starting point in Emmetsburg. Gladfelter is based in Boone at a wildlife research station, and he went out from it to meet the moose. "He was moving right along," Gladfelter reports, "but seemed in good condition and gave no apparent signs of panic or being disoriented. There was a question of what he'd do next, since we are on the edge of the Des Moines metropolitan area."

What the Kid did during the last few days of October was pass through the Des Moines area, an urban center of some 200,000. That, at least, is the only reasonable assumption, because less than a week after he was seen north of Des Moines in Boone by Gladfelter, he was spotted 20 miles or so south of the city, still along the river. Somehow he slipped past the metropolis undetected, a thought-provoking feat for a 1,000-pound animal carrying a four-foot rack of antlers.

"The most amazing thing," says Gladfelter, "is that he must've crossed a couple of interstates and a lot of other roads with heavy traffic." It is possible that some people did see him but didn't believe, or want to report, the evidence of their own senses—a bull moose jogging along I-80.

The Kid made his last appearance in Iowa on the 16th of November, well down the river beyond Ottumwa, more than 75 miles southeast of Des Moines. By mid-December he had traveled another 30 or 40 miles and was below the Iowa line in Clark County, Mo., in the northeast corner of the state. There, another deer biologist, Wayne Porath of the Missouri Department of Conservation, took over as his Boswell.

"I'd heard about the Iowa sightings," says Porath, "so it didn't come as a complete shock when he turned up here. But it certainly was a curiosity. We had an elk wander down from Wyoming along the Missouri River a few years ago. Occasionally roadrunners or armadillo show up in the southwestern part of the state, and we get rumors of mountain lions, but there's never been anything comparable to the moose. So far as I can find in the literature, there's never been a wild moose in Missouri before—at least not since the Ice Age."

However, during the first seven or eight months of 1978 the Kid was seldom spotted in northeastern Missouri, and there was some speculation that he may have made only a cameo, for-the-record appearance. In retrospect it seems likely that he had again found a secluded wooded area (one less accessible to sightseers than had been the case in northern Iowa) and remained close to it through the spring and summer. Whatever his arrangements, he was in fine fettle and full antler by fall, and for the second autumn in a row set off on a grand tour.

Traversing the drainage systems of the Fabius and Salt rivers and Perche Creek, he leisurely circled through north-central Missouri, first heading southwest, then north and finally east, back toward the Mississippi. On this jaunt he was frequently observed, and at one point late in October he was spotted near Harrisburg, 15 miles from Columbia and only about 20 miles north of the Missouri River. It was his most southerly penetration, and by then the Kid had broken all known records for long-distance moose. In Harrisburg he was 100 miles south of the Iowa line and more than 600 miles south of the nearest conventional range for his species on the Minnesota-Canadian border. That's as the crow flies. He had probably wandered twice that distance as the moose walks.

By the first of November he had traveled 20 miles north of Harrisburg and was reported in soybean, winter-wheat and oak-brush country near Moberly, which is the home of Paul Jeffries. Jeffries is a veteran conservation department field-agent who works with area farmers to restore old and create new wildlife habitat. Being intensely interested in the Missouri Kid, Jeffries called Porath, a longtime friend and bowhunting companion, and suggested that they do a little moose looking. The two men didn't catch up with the Kid but spent most of the time following his tracks and deducing from other signs what he had been doing. Mostly he had fed on multiflora roses, which grow in the area, but he had also, fastidiously, without causing much damage, nibbled some winter-wheat shoots. This came as a surprise, because a moose, even if it had reason to suspect that wheat sprouts were tasty, would have difficulty getting at such a low-growing crop. The long, almost giraffe-like legs of a moose give it considerable advantage when it comes to wading and foraging in swamps or reaching up to strip foliage and buds from tree limbs, but the animal is not well designed for bending over and grazing like a sheep or cow. There is just too much moose for this sort of stoop labor.

"That moose figured out the winter-wheat problem without any trouble at all," Jeffries reports admiringly of the Missouri Kid. "He just got down on his prayer bones to get at it."

"On his what?"

"On his knees. The first time in my life I ever tracked an animal who was walking across a field on his knees, but that's what that moose had done!"

Porath had for some time entertained a professional thought that perhaps the Missouri Kid was crazy, there being a parasitical roundworm that attacks hoofed stock, wild and domestic, and eventually reaches the brain. These infestations will often produce aberrant behavior in the host animals. To test this possibility, Porath sent off a bag full of scats to Karns for examination. A few weeks later Karns reported that the laboratory tests were negative. There was no evidence of brain worms, and so far as he could tell from the excrement, the Missouri Kid seemed to be perfectly healthy and sane.

The area around Moberly offers some of the best deer hunting in central Missouri, and because the season was about to open, both Jeffries and Porath were worried that the moose might be gunned down, either by accident or design. Rather than try to guard him or keep his presence secret, they decided that a full-disclosure policy might be best. Through their efforts and those of other state conservation officers, a series of stories was circulated to area newspapers, and radio and TV stations, explaining that there was indeed a moose in the vicinity; that he was a harmless animal of good character and a very rare and interesting one. It was also mentioned that he was not legal game and the law would lean heavily on anyone caught molesting him.

"It worked pretty well," Porath says of the save-the-moose publicity campaign. "We had a lot of calls and letters from people who'd seen him and were pleased and excited about it, and others from people who wanted to know where they could go to see him. I suppose you could say that now he is an object of state pride."

"What it comes down to," says Jeffries, "is that after all the commotion nobody would want to step up and say, "I'm the dirty bastard who shot the Missouri moose.' "

Nevertheless, the Missouri Kid had at least one close call during deer season. Harold Volle is a passionate hunter who lives in the village of Jacksonville. On November 18, he was hunting along Mud Creek, a small stream a few miles to the east of the place where Porath and Jeffries had tracked the Kid as he crossed the winter-wheat field on his prayer bones. "It was kind of a raw day," recalls Volle, "and I left my stand to walk around. All of a sudden I seen something sticking up out of the Mud Creek ravine, some back and a little bitty piece of antler. Now if it had been going away from me, I might have squeezed off a shot. I had the gun up. But it came toward me and got bigger, and I thought there's not a deer alive that's going to stick up that high. I thought, my God, that's the Missouri moose down there in Mud Creek.

"He was only about 100 feet or so away, standing in the rank, old slough grass. He stood about as high as a 15–16-hand horse, and I'd guess maybe he weighed 1,000 or 1,200 pounds. Those big old antlers stuck out three or four feet. He saw me all right, but wasn't a bit scared, just stood there flopping his ears, feeding a little in the slough grass. I must have watched him 30 minutes. Then I took off my hat and started waving at him. To tell you the truth, what I was trying to do was get him to charge, see what he could do. I got myself between two pretty good sized white oaks, and I figured I could dodge around them if he came at me, but he didn't do anything. After a while he just ambled away and I lost sight of him in some oak. It was getting dark and I went home, but I came back the next day with my camera, but I couldn't find him. I almost always carry the camera with me when I'm hunting, but not that day.''

During the next week, perhaps traveling at night, the Missouri Kid loped northeastward unseen through some 15 miles of open fields, mostly corn and bean. On the 25th he turned up near Clarence and was spotted by two hunters who were looking for deer on a farm owned by Everett Johnson. In the winter, when farming is slow, Johnson works as a part-time bartender in a Clarence tavern and from that base added a few comments about the Kid. ''I missed him the day he was at my place,'' he says. ''It was just too cold for me to hunt, but Bud Wirt, who saw him real close, came running up all excited, and Bud has hunted all over, in Montana and Wyoming and other places out west. The moose was wandering around here for a few days, walked around through some cornfields, but nobody minded at this time of year. Besides, he's the most excitement there's been around here since the last tornado.''

As he presumably has in other rural Missouri watering spots, the Kid became a general topic of conversation in the Clarence tavern. One customer brought up a subject that had crossed a good many minds. ''If what you read in the papers is true,'' he delicately introduced the matter, ''that old moose has been a long time without a lady. We may have to start watching our heifers.''

''If the game warden was any good they'd go up north and catch him a lady. Bring her down here for him, and they could settle down in this country. We'd find room for them.''

''That might be no favor for him.''

"How so?"

"Maybe he come down here because he had trouble with his old lady, just said the heck with it and cleared out, started south. He might have no interest at all in seeing any more of old mamma from back up north."

From Clarence, the Kid struck out eastward, following a course that would take him into the rough, tangled, semi-wilderness bottomlands of the Salt River. There were those in the Clarence tavern who worried about what might happen to him there. "Put it this way," one said. "There are some old boys working along the Salt who don't generally let anything get by them alive—in season or out—if you follow my meaning. If that moose gets out of there, he has some kind of a charmed life."

The Missouri Kid had already demonstrated that he was an animal blessed with exceptionally good fortune, and he apparently had no trouble with poachers or anything else in the Salt bottoms. In fact, he seems to have stayed there for almost three weeks before moving on again, this time to the vicinity of Bowling Green, a community 60 miles southeast of Clarence and only 10 miles or so from the Mississippi. There he was observed by Les Brown, a veteran Missouri conservation officer responsible for that district. "There was a lot of talk about what had become of the moose," Brown says of the immediate pre-Christmas period. "It was about the 20th of December. A TV fellow from Quincy [on the Illinois side of the Mississippi] called me and said they certainly would like to get some pictures of the animal, and if I heard where he was, would I let them know. Not a half hour after that, George Linehardt, who has a landfill about a mile from where I live, called and said, "Les, you aren't going to believe it, but I've got a moose up here."

"I went right up and there he was. He looked to be in fine condition. He wasn't spooky at all, but it's open up there and he didn't stay long. He trotted off toward the woods. On the way, he sailed over a fence without any trouble. He just tucked up his front feet like a jumping horse and cleared it with no struggle."

"How high was the fence?"

"It was four feet with two strands of barbwire on top."

Jumping such a fence may not be an especially impressive physical feat for one of the Kid's size and build, but behaviorally it gives pause for thought. A fence is not a complicated device, but in the subboreal wilderness that is their customary home, moose don't have to cope with such flimsy appearing but dangerous barriers. Somewhere en route the Kid met his first fence and had to learn something of its properties and what to do about them. Along with dealing with multiflora roses, winter wheat and interstate highways, fence management is now one of his known acquired skills. He may well have a good many others that nobody has yet seen displayed. It's quite possible, because of the experiences he has had and the adaptations he has had to make, that the Missouri Kid is the best-educated as well as the most-traveled moose in the world.

Bowling Green sits on something of a ridge. Between the ridge and the Mississippi there is a complex of bluffs, ravines, relatively thick woods, brush and swamp that parallel the river for 30 miles from Hannibal on the north to another old steamboat port, called Louisiana, on the south. It's as similar to traditional moose habitat as anything there is in Missouri, and the Kid apparently spent the holiday season in this wet bush. For dramatic as well as ecological reasons it was a very appropriate place. The Hannibal-Louisiana area, much of which lies in Pike County, is special enough to warrant a digression, though given the peculiarity of the Kid's case, the area may be a crucial factor in his story and discussing it not a digression at all. Getting to the Hannibal-Louisiana part of the world might be the mysterious purpose of the Kid's great journey, a purpose perhaps conceived by something beyond a moose's instincts.

As much as any other real place, this Pike County country is Raintree County, U.S.A. One former resident, a fellow by the name of Sam Clemens, who was born a few miles up the Salt River, came closer than anyone else to finding the mythic heart of our land in these parts. Ever since, it has been a sort of American Logres, a place where reality and illusion shift and converge like bits of glass in a kaleidoscope to form new patterns that are more provocative than fiction and more instructive than fact. As one gross example: for a century and more this nation has reported sub- or super-humanoids—Bigfeet, Sasquatches, Wild Men.

Mark Twain, which is the fancy name Sam Clemens gave himself after he began to journalize in and around Hannibal, Louisiana and Pike County, claimed he had located and interrogated one of these creatures. It was a surprisingly easy interview because, the Wild Man told Twain, he'd been waiting for some time to give his life story to an open-minded newsman.

The Wild Man claimed to be the son of Cain, but, he said, "In these degenerate days I am become the slave of quack doctors and newspapers. I am driven from pillar to post . . . at the behest of some driving journal. I am bundled off to this howling wilderness to strip, and jibber, and be ugly and hairy, and pull down fences and waylay sheep, and waltz around with a club, all to gratify the whim of a bedlam of crazy newspaper scribblers."

After having ascertained that the Wild Man had not "given these items to any other journal," Twain, the complete professional as always, made sure he had the name right. The Wild Man said it was Sensation.

"All of which is in strict accordance with the facts," Twain concluded in his report.

Ever since, there have been many Sensations in the area. The Wild Man, or his kin, revisited Pike County in 1878, 1908 and 1963. In 1972 he became (perhaps under the influence of flying saucers and fireballs, which were also common that summer in Pike County) the Missouri Monster—or, as he was familiarly known, MoMo. He was first sighted by two Louisiana schoolchildren, who said he resembled a shaggy overgrown ape that smelled like a dead horse. Local residents were bothered not only by MoMo, but also by hordes of newspaper, radio and TV people, as well as by UFO investigatory committees. None got an interview with MoMo, which suggests that journalists as well as monsters aren't what they were when Twain had this Missouri beat. The 1972 Sensation was considerable and lasted until the end of July, when MoMo was apparently assigned elsewhere.

The point is that people in and around Pike County have been sensitized to the extraordinary, have a long tradition of dealing with it easily and imaginatively. If there is any one place that deserved to have a real moose who had made an all but mythic journey, this is it.

When the Missouri Kid started traveling again early this year after having spent a quiet Christmastide in the bottomland boondocks, he appeared first to a deserving Louisiana couple, Rich and Donna Lord, who live on a wooded knoll a few miles from Louisiana. Donna Lord was especially pleased to see the Kid because in 1972 she had missed meeting MoMo. "I think there may have been something out there," she says in regard to the Missouri Monster. "I know a woman who isn't a radical—in fact, she's a very religious woman. She said something came up and shook her trailer. The stink was awful, and her phone didn't work until it went away."

In the case of the moose, there was no room for doubt. On January 25, Rich Lord, who had been cutting firewood, came running into the house to tell his wife that the moose was in an adjacent field. The Lords called the *Louisiana Press-Journal,* a publication that, according to local tradition, once employed Twain. Very quickly about 50 people collected near the Lords' house. "The state police were directing traffic," recalls Fred Burgess of the *Press-Journal* editorial staff. Among those in the crowd was a local dentist and amateur photographer named Dr. Frank Thomalla, who had left a patient sitting in a dental chair.

The Lords volunteered to drive the Kid out of the woods for the benefit of the media and spectators. This sort of thing is Donna Lord's meat. "I don't hunt, because I'm scared of guns," she says, "but I love to chase things. I'll go out on a deer trail until I run it out. My boys always want me to go with them when they hunt because they say, 'Mom always finds us a deer.'"

The moose presented no problems for an experienced beater like Donna Lord. Obligingly, the Kid allowed himself to be urged across a nearby paved road to an open area where the *Press-Journal* staffers and Dr. Thomalla photographed and admired him. Later the Lords repeated the drive for the benefit of a late-arriving TV crew.

Perhaps feeling he had done enough for the media, the Kid next made several educational appearances, showing up a day later at the Boncl elementary school some five miles south of the Lords' property. "We were dismissing, waiting for the bus," says Marcus Yelverton, the Boncl principal. "I happened to mention that there was supposed to be a moose in the area. About three minutes later there he came. He walked slowly across the field right in front of the school. He was a beautiful animal. Probably when they're old and have forgotten my name and most of their classmates, those kids will remember the day when a moose came to their school."

The Kid kept to himself over the weekend, but by Monday he had doubled back north toward the town of Louisiana. Crossing a highway, he paused and displayed himself so as to provide a memorable moment for a speech teacher and two of her students, who were traveling to an oratorical contest.

Donna Lord, by reason of her first-hand experience with the moose, has developed a proprietary and protective interest in the Kid. "I can't think why anybody would want to do him any harm," she says. "In the first place, he's been feeding on brush, and if you were to try to cook him, you'd probably have to clear out of the kitchen for two days or so. But more than that, he's come such a long ways and done so many interesting things, it would be awful to kill him. What worries me is that weird people hear about something like the moose and they get strange ideas, decide they want to do something to him, maybe for the publicity."

Unfortunately, as Donna Lord suggests, celebrities do attract crazies, and the Missouri Kid has done so. "It isn't generally known," says Jim Schwartz, a young conservation officer stationed in Louisiana, "but one of our agents in the St. Louis office got a tip from one of his informants that the moose had been killed. He gave the date and place and even the time of day."

"An informant?"

"All law agencies have them."

"What happened?"

"We checked it out and didn't find anything. Then a couple of days later I had a good report of the moose crossing Highway 54. As far as we know, the story was false."

"Why would anybody want to turn in a tip like that?"

"Maybe the informant was just testing our agent to see what his reaction would be."

According to Schwartz, the Kid was last seen in the flesh about mid-

February when he crossed a state highway heading into the roundwood country of northern Pike County. However, there was later circumstantial evidence, as incontestable as a trout in a milk pail, as to his subsequent whereabouts and activities. About six weeks after the last sighting, Marion Traynor, the chief operator of the Louisiana waterworks, was scouting around in the woods in preparation for the opening of the spring season on turkey gobblers, birds he loves to hunt with a muzzle-loading black-powder shotgun. In a thicket near an old quarry northeast of Bowling Green, Traynor came upon a set of moose antlers, the halves of which were lying within 10 feet of each other. "I have a little schooling in conservation and I own a lot of outdoor books," said Traynor. "So with all the stories of this moose, and knowing from reading what they should look like, I knew what these were right off. There were signs he had been lying around in that spot for a few days."

Traynor measured the antlers (they were 41 inches tip to tip with a seven-inch skull space) and then took them to Jeff Pennock, another state wildlife biologist who resides in the area.

There the matter rests for the moment. The consensus among mooseologists is that the Kid is in the ravine and brush country that covers much of the area between Louisiana and Hannibal — lying doggo as he did during his antlerless periods in Iowa in 1977 and in northeast Missouri in 1978.

"There are places in those bottoms that we call Africa," says Brown, the veteran game warden who knows the country as well as anyone. "A moose or anything else could lay up in there for a long time without being seen."

That is a reasonable assumption. Beyond it there remains considerable speculation about what the Missouri Kid will do and where he'll go next, say in the fall. One fanciful theory is that his ultimate objective is to get to New Orleans for Mardi Gras and that he was just waiting around in Pike County for the police strike to be settled in that city. While this might seem beyond the realm of possibility, the Kid has already stretched a good many realms beyond what had previously been regarded as possible. In a technical zoological way, he has expanded the known range of his species farther than any other moose. Indisputably he has enlarged the folklore of the true-blue classic American Sensation. Even more significantly, if one cares to pursue the subject, the Missouri Kid has greatly deepened the always intriguing mystery that has to do with the inner world of other species.

Addendum

Since this account was written I have been in Missouri frequently, in fact lived there for a year, and have made regular inquiries about the moose. There have apparently been no further signs or sightings of the animal. State game agents who share this interest find the absence of reports to be almost as thought-provoking as the appearance of the animal. They reason that since moose remains would be so large and unusual, they would have received word if the Missouri Kid met with an accident—say on a highway—died of natural causes or even if he had been taken by poachers, who tend to be constitutionally unable not to brag about their feats. It is also reasonable that if he moved on to another, normally mooseless state, stayed there for any length of time or met his end there, the animal would have been discovered, alive or dead, and made news. Therefore, it is at least conceivable that The Kid headed north in the spring of 1979 and, perhaps following the brushy Mississippi bottoms, eventually returned to Minnesota to become again an unremarkable member of the resident moose community. There is no precedent for a moose making or being able to make such an extraordinary journey, but again there is no hard evidence that one could—and has—not.

An Eyeshine of Ferrets

For a period in the mid-1960s I was a journalistic observer of and commentator on the federal endangered-species program that was then being organized to aid animals thought to be in imminent danger of extinction. Thus I happened to spend a few cold days in the fall of 1967 in western South Dakota with a field biologist named Don Fortenberry, who was at the time 35 years old. He had been assigned by the U.S. Fish and Wildlife Service to find and learn something about the black-footed ferret, which at the time was regarded as the least numerous and thus most endangered mammal in the United States. It still holds that dubious distinction.

There were—and are—more assumptions than facts available about the black-footed ferret, a member of the weasel family. Thirteen years ago one guess was that there were no more than one or two hundred of the animals in the United States. They had once inhabited, always in sparse numbers, most of the Great Plains, and a few could still have been living anywhere in that vast region. However, most reports of their continuing existence came from the prairie and butte country of South Dakota west of the Missouri River. Even if the ferret had been numerous there, Fortenberry would have had a formidable problem spotting any, this being a very elusive and reclusive beast. Invariably associated with prairie dogs, the ferrets conduct most of their activities, predatory and otherwise, underground in the mazelike burrows dug by those communal rodents. Also, ferrets are nocturnal and seldom surface except in the dark.

The scarcity of ferrets further complicated Fortenberry's work. On the basis that the baby shouldn't be thrown out with the bath water, he had to forgo many standard search techniques. Running trap lines through prairie-dog towns might have produced a few specimens — weasels, as a family, being easy to snare — but there was the worry that ferrets thus killed might be the last ones left.

Fortenberry (and others who have followed him as ferret researchers) therefore proceeded cautiously. By day he would examine prairie-dog towns for surface signs of ferrets. But the ground in this region is hard, doesn't take good impressions and, therefore, is poorly suited for sign reading. Such signs as are made don't last long because of the scouring effect of the prairie wind. Also, prairie dogs are notable excavators, forever rearranging the earth in the vicinity of their tunnels, and that effaces signs. So Fortenberry depended principally on nighttime searching. He would take a four-wheel-drive vehicle to a spot with a good view of a prairie-dog town and from time to time through the night sweep it with a searchlight on the chance that he might spot a ferret, or at least its eyeshine, a brief spark of peculiar greenish reflection. Among animals customarily found in prairie-dog towns, only the eyeshine of the long-tailed weasel is similar to that of the ferret.

On the last night that I went searching with Fortenberry, we set up about dusk on the property of an obliging rancher. The prairie-dog town we were watching covered about 75 acres, extending from a gulch toward a solitary butte under which we parked. It was a dark, overcast November night, without moon or stars. A sharp wind out of the northwest rattled the dry prairie weeds and drove before it a lot of gritty dust and a few grains of snow almost as abrasive. Huddled in the cab of the truck, bundled in goose-hunting clothes, we drank coffee and talked about ferrets, politics in the Interior Department, world affairs, ball games and other things two men might be expected to discuss when they have to sit up all night. Every five minutes or so Fortenberry would play the spotlight across the dog town. Quite often it caught something. Jackrabbits, because of distortion caused by the light and distance, looked pale and as big as cocker spaniels. Two coyotes, a raccoon and a yellow plains porcupine of vaguely prehistoric appearance stood at different times transfixed by the beam. I found all this more interesting than did Fortenberry, who had already seen perhaps too much of prairie night life. Nevertheless, each time the light went on he would strain forward toward the windshield in anticipation. It was a compulsive reaction illustrating the power of faint hope over high probability—rather like casually buying a 25¢ chance in a million-dollar lottery but getting edgy on the day of the drawing.

Just before dawn we both thought we saw a suggestive glint of reflected light, but it disappeared before either of us could speak. Fortenberry worked the area over and over with the beam, and after the sky became light we went out and searched the ground for signs. There were none, and Fortenberry said, "It's easy to see spooks when you do this work." We talked about things mystics say they have seen in a single point of flame.

The flick of green light may have been the reflection of something more than wishful imagining, because Fortenberry did find a ferret later on not far from where we had spent that night. I was long gone by then.

My ferret-hunting expedition was a classic non-event. However, I have thought often of that night, and I think I recall it more clearly than I do many other more conventionally eventful ones. Certainly, the trophy-hunting possibility helped make it memorable—the faint chance of seeing, and thus figuratively claiming, something of extreme rarity. Beyond that, there was a powerful surrealistic quality to that night as if some elemental force was floating around in the dust and snow, mixed in with pale rabbits and yellow porcupines, a force having less to do with ferrets than with men—the two of us and a good many others of our species whose interests Fortenberry and I, in a way, represented.

Though my direct involvement with endangered species subsequently declined, I remained interested in them, and particularly in the ferret. During the 1970s, when things seemed to be looking up for many of the hard-pressed animals—the whooping crane, the masked bobwhite, the everglade kite, the sea otter, the eastern timber wolf—there never was good news, or much news of any sort, about the ferret. Ecological and political problems relating to the animal seemed to grow more complicated. Some months ago, I learned through reports and conversations that the federal agencies had come to an administrative and biological dead end with the species. This suggested that if anyone wanted to ask questions or say anything about this elusive mammal in any but purely historical terms, it might be well to do so soon. So 13 years after my night on the prairie with Don Fortenberry, I returned, in a sense, to the ferret.

Evolutionists generally agree that the ferret, as a distinct member of the weasel family (the *Mustelidae*—a clan that includes skunks, badgers, minks, otters, wolverines and a lot of lesser beasts called simply "weasels"), originated in the Mediterranean basin. They are lithe, elongated carnivores especially well equipped to pursue tunnel-dwelling rodents—in the case of the black-footed ferret, those rodents are prairie dogs, who not only build the ferrets' homes for them but serve as their dinner as well. Thousands of years ago these abilities came to the attention of human beings in the Old World who caught some ferrets and ever since have bred and kept them as hunting aids, particularly for rousting rabbits out of their burrows. The descendants of these animals, as thoroughly domesticated as the dog or cat, are known as European ferrets. They are common in this country as laboratory animals and even as household pets.

Other ferrets in ancient times began an immense eastward migration. Some eventually arrived in the steppes of central Asia and remain there today as a feral species called the Siberian ferret. Others continued to pioneer, coming by and by to Alaska, presumably getting there over the Bering Sea land bridge that once linked the Americas to Asia. These were the ancestors of the black-footed ferret—*Mustela nigripes.* Somewhat like the Indians and Eskimos who got to North America by the same route, *M. nigripes* is such an ancient resident as to be considered a native of the continent. However, except for a dark facial mask and the four black feet that gave it its popular name, the animal isn't much different physically or in its behavior from the wild Siberian or domesticated European ferrets. Like them, it reaches a maximum weight of around 1½ pounds, and an adult is between 18 and 22 inches long, including its 4–5" tail.

Rounding the Arctic corner, the ferrets, according to their fossil remains, continued south and finally settled down in the Great Plains. Being minor predators, they probably were never very numerous, and they have certainly never been conspicuous so far as humans are concerned. Except for a few tiny mouse-shrewlike creatures, ferrets were the last mammals in the U.S. to be discovered by naturalists. They received their first mention in 1851 after John James Audubon had received a single skin from a collector in Wyoming. The Plains Indians traditionally used bits of ferret skin in ceremonial garments; this use suggests that the animals were always rare, or at least rarely met by people.

Ferrets have been seen, found in traps or discovered as road kills only occasionally since Audubon's day. Increasingly, reports of their existence were restricted to South Dakota, where since 1889 there have been about 400 valid sightings—an average of four or five a year, although the rate has been lower in recent years. The last bona fide recorded encounter with a ferret occurred more than a year ago in Todd County, S. Dak., when one was met by Dennis Lengkeek, a state conservation officer who has worked closely with the federal ferret researchers.

Although no one can currently produce a living ferret or do more than hazard a guess as to where one might be, nearly all experts on the animal believe—have a "gut" feeling, as it is often expressed—that there are still some ferrets, most likely in South Dakota, with Wyoming another strong possibility. It is also accepted that the range of the animal has been drastically reduced since Audubon described it 130 years ago, and there isn't much debate about the principal reason for the decline.

When white settlers began to occupy the Great Plains, one of their first projects was getting rid of prairie dogs. There were a lot of them, perhaps as many as five billion, and in some places contiguous prairie-dog towns occupied as much as 25,000 square miles. The burrow entrances of these animals would occasionally break a horse's leg or a wagon axle, but, more important, cattlemen came to the conclusion that these browsing rodents were formidable competitors with cattle for forage. This remains a matter of hot dispute in ranch country. There are studies—the most recent was released this year by the U.S. Forest Service—that indicate prairie-dog damage may not be as great as suspected. This finding is supported by a historical argument: at one time the plains, occupied by billions of prairie dogs, also supported 60 million buffalo and 40 million antelope, both notable grass eaters.

Whatever the ecological truth, most ranchers became convinced that prairie dogs were economically intolerable and had to go, or at least be thinned out considerably. Private landowners asked for and received public assistance in doing the chore. The states—and, after 1914, the federal government—got into the business of controlling prairie dogs as a form of agricultural protectionism.

"Control" is a bureaucratic euphemism that means kill. In the case of prairie dogs, it means poison. Public agencies had at the rodents with a variety of lethal gases, powders and solutions, either to eliminate colonies or to keep the area they occupied very much reduced. Nobody knows how many prairie dogs were controlled, but the body count was certainly in the millions. By the middle of this century the animals had become locally extinct in many areas of their former range.

Neither ranchers nor government exterminators had anything against the black-footed ferret: in most cases they probably didn't even know the beast existed. However, the circumstantial evidence is overwhelming that prairie-dog control played havoc with the ferrets. They succumbed directly because of the poisons, and later, as the dog towns were eliminated or reduced, the ferrets died off because of the loss of food sources and effective habitat. As early as 1929 Ernest Thompson Seton, the most influential popular naturalist in the U.S. at the time, was predicting that the black-footed ferret was not long for this world if massive prairie-dog poisoning programs continued. However, it was hard to arouse much public indignation about the plight of an animal few people had ever heard of, much less seen or admired.

In the early 1960s the situation changed dramatically. Alarmed by, among other things, Rachel Carson's *The Silent Spring,* environmentalists began attacking the unchecked spread of pesticides, herbicides and commercial and industrial toxins. The work of public wildlife exterminators was cited as particularly outrageous. For example, in one year in the 1960s the federal Fish and Wildlife Service had controlled, mostly by poison, 154,640 wild animals, including 77,000 coyotes, 850 grizzly bears and 280 mountain lions. Not only because of the amount of controlling it was doing but also because it was supposed to be the national protector of wildlife, the agency caught a lot of serious heat.

Environmentalists considered one poison especially objectionable. It was a sodium fluoroacetate called Compound 1080, which was not only very lethal but also an "indiscriminate" toxin, as opposed to a "specific" poison. Compound 1080 slaughtered a lot of animals other than those for which it was intended. A prairie dog that had eaten 1080-treated grain became a secondary bait for anything that ate it. It was argued that 1080 was killing off uncounted and uncountable numbers of creatures and would bring about the quick extinction of at least one—the black-footed ferret. The fate of the obscure ferret thus became something of a symbolic rallying point in the overall antipoison campaign.

Fanning the public outrage was the fact that in 1964 the U.S. Fish and Wildlife Service began to compile a list of endangered species, creatures who were now so few in number that their survival was in question. The ferret headed the list of endangered mammals. Environmentalists and ferret students immediately raised the obvious point—that it was hypocritical, to say the least, for one branch of the Fish and Wildlife Service to declare the ferret endangered while another branch of the same agency was busy stuffing poisoned grain down prairie-dog holes.

South Dakota became the focal point of this controversy. A likely reason why people had continued to see a few ferrets in South Dakota was that for a long time there had been less prairie-dog control in that state than elsewhere, because much of the western part of South Dakota was given over to Indian reservations. After World War II, however, Indian ranchers demanded the same kind of prairie-dog control their white counterparts had long enjoyed. A lot of the dogs were subsequently controlled in South Dakota.

Environmentalists insisted that these operations should cease or be curtailed, generally because poisoning was bad, and specifically to save the ferret. Ranchers, the Bureau of Indian Affairs and state and federal exterminating agencies heatedly rejoined that while they wished the ferret well, it was unfair to force them to give up poisoning prairie dogs—and suffer the economic consequences of doing so—just because some animal of no known value that nobody ever saw might be accidentally harmed.

Several years of bitter wrangling and bureaucratic infighting followed. In 1972, in what was regarded as a considerable triumph for environmentalists, President Nixon signed an order forbidding the use of 1080 by anyone—feds, state agencies or private parties.

After the 1080 ban the federal government got out of prairie-dog poisoning in South Dakota. In fact, very little control work was done there by anyone for five years. But in 1978 a new poison, zinc phosphide, was allowed to be used by the state of South Dakota. It effectively kills prairie dogs; however, it is more time-consuming to use and thus more expensive than 1080 was. As a result, the South Dakota prairie-dog population has indisputably increased in recent years, and there is agitation to go back to the good old days of 1080.

Earlier, however, while attempting to settle the prairie-dog poisoning dispute, the Fish and Wildlife Service organized a project designed to find some ferrets and develop ways to protect and increase the population. That was why Don Fortenberry had been sent to South Dakota to do field research in 1960. Conrad Hillman, a graduate student from the University of South Dakota, was also working on ferrets at the time, and the two cooperated closely until 1972, when Fortenberry was reassigned to make an impact study of the proposed Alaska "Wilderness Lands" bill. Hillman then became the federal ferret man, the only one in the world.

By checking reports from private landowners and state game officials and spending thousands of hours looking for ferret signs, Fortenberry and Hillman had been able to locate 71 of the animals during the first eight years of their study. Included were 38 young ferrets found in 11 litters. This was a larger number than anyone had thought would be found, and in 1971 plans were made to live-trap some of the animals in the hope of breeding them in captivity at the endangered-species laboratory the Fish and Wildlife Service maintains in Patuxent, Md. The endangered-species specialists had had considerable success breeding such rare creatures as whooping cranes and masked bobwhites, and because weasels in general (for example, the mink) breed well in captivity, it was assumed the black-footed ferrets would do likewise.

Fortenberry and Hillman therefore trapped six ferrets, four of them females, in Mellette County, South Dakota. The plan was to hold the animals for several weeks of acclimatization in South Dakota and then fly them to Patuxent. However, before they ever got to Maryland all four females died. Reports from the Fish and Wildlife Service say only that the animals died. That the ferrets' fate is dealt with in such extreme brevity is understandable in light of what occurred.

"What happened was that we were responsible for the deaths of those animals," says Hillman. "It was decided that the ferrets should be inoculated against distemper. Everyone agreed. There were two choices of vaccine. We thought and talked about it a lot—and we made the wrong choice. We sent the bodies to Cornell University for examination. The report was, essentially, that we had given them a fatal dose of distemper."

Despite this disaster, the two surviving males were sent to Patuxent in November 1971, where they were later joined by a female, trapped in 1972, and a pair, presumably mates, taken in 1973. Colonies of European and Siberian ferrets were also established at the laboratory to serve as surrogate study animals.

The five black-footed ferrets adjusted well enough to captivity but they did not, to put it mildly, reproduce vigorously. The single female, nicknamed Frigid Min, never became sexually responsive. Close examination of the pair trapped in 1973 indicated that both were well into middle age—10 years or so being regarded as a good life-span for ferrets. The female was thought to be perhaps eight years old and was called Granny. Nevertheless, she conceived in 1976 and had a litter of five. But four were stillborn and the fifth, toward which Granny hardly behaved in a maternal manner, was weak and died within two days.

Granny bred again the next year, and as whelping time neared, the laboratory staff set up a round-the-clock watch. Again five young were born. Again four of them were stillborn. The fifth infant was removed from Granny, artificially warmed, fed, carefully doctored. It was offered to a nursing Siberian ferret, which willingly accepted it—to no avail. This youngster also died two days after birth. By 1978 Granny was truly an ancient ferret, 12 or 13 years old. But she was the only female available and so she was bred once more. Conception apparently occurred, and it was decided to remove the young by cesarean section. The operation was performed, but it turned out that Granny had had a pseudopregnancy.

A few months later, in January 1979, Granny died of mammary-gland cancer. She was the last of the black-footed ferrets in captivity, her former mate and the other animals having succumbed earlier to other cancers.

Jim Carpenter, a Patuxent research veterinarian who attended the ferrets, is of the opinion that the cancers (which didn't affect the Siberian ferrets), as well as diabetes and some other congenital ailments that afflicted the blackfoots, indicated a pattern of genetic weakness that may be influencing the fate of wild ferrets. "The wild population may be so reduced and so isolated that inbreeding has become a problem," Carpenter says.

However, Carpenter feels that the Patuxent ferrets did not die in vain, because the laboratory's experience with them provided a great deal of information, especially about the animal's reproductive physiology. He believes that, given stronger stock, ferrets can successfully breed in captivity. That may be the last and the best hope for preserving the species.

Meanwhile, back in the field, the ferret research that had begun so promisingly also started to encounter afflictions of a terminal sort. The number of sightings of the animals declined after 1974. Though there continue to be a few valid reports, Hillman, who has spent more time than anyone else looking for ferrets, has now not seen one in five years. Eighteen months ago, in an attempt to do something about the searchers' lack of success, the Fish and Wildlife Service contracted for the training of four dogs as ferret hunters. The dogs were conditioned to respond to the scent of ferret scats sent from the Patuxent laboratory and two of the dogs—both Labrador retrievers—proved in training exercises to have considerable aptitude for this work. However, the dogs, too, were unsuccessful in their efforts to find wild black-footed ferrets.

The paucity of sightings may mean that the worst possibility has come to pass—that the species is now balanced on the brink of extinction or may even have fallen into the eternal abyss. "Anything is possible," concedes Hillman, "because there is so little that can be objectively proved. Maybe there is some critical number below which the ferrets cannot reproduce effectively. It could be that the family groups are so small and scattered that vigorous males and females cannot find each other easily or safely."

However, Hillman, as well as other biologists who have worked with ferrets, is more optimistic than that statement indicates. "My gut feeling is that there may be as many ferrets as there were when we started looking in the 1960s," he says. "There is no question that after the 1080 ban, prairie-dog colonies have expanded. There may be two million acres of them in the northern Great Plains now. Consequently, there is more ferret habitat than there was. This in itself may be a factor in the lack of sightings. Ten years ago I'd spend a couple of days going over a 50-acre dog town looking for signs. The same dog town may be three times as big now, which means it takes a lot more time to look it over and the chances of missing something are at least tripled.

"There is another thing about the lack of recent observations. I'm the ferret man, and *I* haven't looked as hard during the last few years as I did before. I've been doing other research, too, with swift foxes [another rare prairie species], for instance—at least you occasionally see *them*. When I was in my 20s I'd stay in the field for weeks on end, looking for ferrets day and night. One fall I lost 30 pounds. It wears you down, physically and psychologically. Then, too, I was with Fortenberry much of the time. You know what just a few days of looking for something you almost never see and can't really prove is even there does to your head? Fourteen years of it gets to you."

The lack of ferrets in the field and the collapse of the captive breeding program have also frustrated Fish and Wildlife administrators in Washington. Unlike their efforts with the whooping crane and the wolf, the ferret project has given the feds nothing to brag about, no scientific or political kudos. And because some $600,000 has been spent on this obscure animal, the program becomes increasingly vulnerable to Golden Fleece criticism—in short, that it is a foolish waste of public money. In 1979 a story saying just about that appeared in the *Wall Street Journal*. Besides taking a few gratuitous swipes at some of the people involved, the report was inaccurate—among other bits of misinformation, the *Journal* erroneously stated that $20,000 had been spent to import the dung of Siberian ferrets. Even so, it was the sort of publicity that makes bureaucrats very nervous.

Though the feds insist the *Journal* article was not a matter of concern, funding for further ferret research has been withdrawn from the proposed 1981 Fish and Wildlife Service budget. That was discovered in January of this year by a seven-member organization called the Ferret Recovery Team. (Several endangered species have teams of government and private experts who have a particular interest in the species and provide advice for its welfare. The Ferret Recovery Team is just such a group.)

In February, to satisfy my curiosity, I visited with a group of Fish and Wildlife Service administrators who assembled in a Washington office to explain the decision to stop ferret research. The chief spokesman was Harold O'Connor, deputy associate director of federal assistance, a position that puts him second in command of endangered-species programs and thus makes him one of the architects of ferret policies. Much of the conversation was semantic, having to do with what word or phrase best described what had happened to the ferret program. The Fish and Wildlife people thought that expressions like "end," "abandon" and "close down" were misleading, despite the fact that their budget proposal had used the word "terminate." Glen Smart, an endangered-species staff research specialist, suggested "de-emphasize." This was accepted as the best and most accurate word.

Ferret de-emphasis was then explained. Only $66,000 had been allotted for ferrets for 1980, because funds were very tight—there was only about $1 million available for all endangered-species research. Ferret research *had* to be deemphasized. Previous investigation had turned up a lot of valuable information, and it now seemed time to phase out that part of the program. De-emphasize.

"But the fact is that after this budget cut you won't have money for a ferret man," I suggested. "You won't really be in the business any more."

No, that wasn't really so. All over the land Fish and Wildlife agents would be thinking about ferrets, would be ready to protect any ferrets somebody else located. Hillman would be reassigned to work with wolves in Minnesota, but the service would still have his expertise and he could be sent back to South Dakota in a matter of hours if there were any hot ferret leads.

All this seemed reasonable, if not ideal, until later, when a long-time service friend said informally, "You probably didn't know, but Hillman has resigned. He's leaving as of March 1."

"What was all that about his being reassigned and being held in a state of ready-alert to do ferret work?"

"I don't know, but he's quit. Not just the ferret job but the whole service. He is going to do research for a private conservation outfit."

Federal officials, even when one is inclined in their favor, are sometimes hard to love.

Hillman, in his final week as the last of the federal ferret men, was not at all bitter, but he was outspoken. "I could see this coming," he said, "Washington getting restless about their commitment to the species. I'd still be here if I thought they were really going to back it, but I didn't want to stay on as a kind of token. Obviously I care about these animals. I'm going to stay in touch and maybe I can do more from the outside than I could on the inside."

I asked him what he would do if he had his hand on the keys to things—the authority and the money.

"The first thing is to find ferrets," he answered. "I think they are there. The dogs are not a bad idea, but mostly we need bodies. One man can't do it. I don't mean there have to be a lot of full-time employees, a big budget. There are ways we could get help—other agencies, maybe some graduate students, volunteers who would put in some time. I talked to officials from the Navaho reservation in Arizona. They're interested in ferrets and they may have some—nobody has ever really looked there. The Navahos might put some time into looking for them. If we turned up some ferrets I don't think it would be hard to make arrangements to protect them, either on federal or private lands."

"What about trying captive breeding again?"

"I'm not so sure about that now. Sometimes I say that if I found a ferret the best thing I could do for it is not tell anyone, but that's easy to take out of context. What I mean is, I think the first priority is to find out a lot more than we know now about the natural history of the animal, the reproductive, territorial, predatory requirements, how the litters disperse, what would make a reasonable sanctuary area. I think good, continuing field observations are what we need most. Then we would stand a much better chance of breeding animals in the lab and transplanting young ones back into the wild."

According to the Washington administrators, the main ferret man, now that Hillman is gone, is Maurice Anderson, a veteran wildlife biologist who is an endangered-species specialist assigned to a branch office of the Fish and Wildlife Service in Pierre, South Dakota. Anderson is responsible for filing reports, providing public information and sitting in on meetings about a variety of scarce species found in the northern Great Plains. As part of that job he was appointed to the Ferret Recovery Team, but he makes no claim about having special expertise with the species. "I've picked up Con Hillman's papers," says Anderson, "and I'll go on filing ferret reports, if there are any."

"What would you do if, say, tomorrow you got what sounded like a very good report?"

"I'd try to arrange to get out of the office and go look," says Anderson, a soft-spoken, low-key, commonsensical man. "If there were good signs maybe I could get the dogs and check them out."

The two Labradors are in effect the last full-time federal ferret staffers.

"If they did check out, if you actually found a ferret—what then?"

"There wouldn't be a lot I could do immediately. We don't have funds or personnel for research. I'd go through the chain of command. What happened would be a Washington decision, I suppose."

Ed Brigham is the director of regional activities of the National Audubon Society. Until last year he was a member of the Ferret Recovery Team, the only one who wasn't a public employee. "Whatever they call it—terminate or deemphasize—I think the Fish and Wildlife Service has made a bad mistake," he says. "The ferret is the first major endangered species they have given up on. Research possibilities do remain—they were spelled out in the recovery plan we submitted to the service in the summer of 1978—but nothing much was done to implement them. You get the feeling that people in Washington are interested in things that promise quicker results and better publicity. I'm afraid other agencies and other people—say, those pushing to use 1080 again—are going to take this as a sign that there is no good reason to pay much attention to this animal any longer. I have to think that the ferret has suddenly become much more endangered than it was before this decision was made."

Having terminated and thus de-emphasized conversations with public officials, I drove out to the South Dakota countryside looking for a prairie-dog town where, according to Anderson, a ferret *might* have been seen a few years ago. The wind was again whipping snow and dust across the prairie, but not even in a 25¢-against-a-million-dollar-jackpot way was this a ferret-hunting expedition. It was simply that a prairie-dog town provided a better environment for thinking about ferrets than federal office buildings and laboratories do.

When, 13 years before, Fortenberry and I had sat in a similar place, we had talked in a self-mocking way about what a conventionally worthless thing a black-footed ferret is. They make up such a minuscule portion of the animal kingdom that they are of almost no ecological consequence, even to prairie dogs. If, as some environmentalists think, protecting the diversity of the world's gene pool is important, ferret genes are in good supply from the plentiful and almost identical European and Siberian species. So far as humans are concerned, we've never, except for the Sioux robe makers, had any practical use for the ferret. We have known it so briefly and imperfectly that it isn't a creature with historic or legendary associations for us; it doesn't conjure up atavistic remembrances of things past, the way the wolf does. Ferrets have never pleased or stimulated us esthetically, as the whooping crane has, and they probably never will, because they are more or less invisible. Fortenberry and I discussed all of this, but we couldn't get around the fact that we were where we were that night and that others had and would be in the same sort of improbable place for the same insubstantial purpose.

It would be tidy, but sheer contrivance, to claim that 13 years later, while kicking frozen clods in an empty prairie-dog town, these paradoxical matters sorted themselves out in a blinding flash of insight. About all that did occur to me was the notion that it wasn't a bad way to spend a few hours, that I had spent some of the best parts of my life looking, in a sense, for ferrets. So has everyone else I know.

A very common, in fact almost definitive, characteristic of our species is that we often become passionately concerned about things of no intrinsic value. Caring deeply about paintings and houseplants, honor and free speech, the outcome of ball games is not unlike caring about ferrets. Individually—and collectively—we have a lot of Golden Fleece interests. We can, if circumstances require, sacrifice some of them without much suffering, but at the same time we know that without some of them life would be brutal and almost unbearable.

Perhaps we are now as a nation too poor to continue a public search-and-rescue operation for ferrets. If so, the general quality of life won't be endangered. In fact, if the last ferret should shuffle off this mortal coil (or already has), there will be no practical reverberations. Yet there are real limits to how much of this sort of cost-accounting we can afford. The cost, as well as the considerable glory of being human, is that now and then we must go out into prairie-dog towns and look for ferrets. No ferret will ever come looking for us.

Addendum

Mr. and Mrs. John R. Hogg are Wyoming ranchers. Their property is located in the northwestern part of the state, near the village of Meeteetse, where the Great Plains begin to rise into the Rockies. On the evening of September 25, 1981, the Hoggs heard their dog Shep barking outside the ranch house but did not think that the commotion required investigation. The next morning they found that Shep had killed but not badly mutilated a vaguely minkish animal, unfamiliar to either of the Hoggs.

Susie Hogg also operates a cafe in Meeteetse and the next morning when she left for work, she took the odd remains with her and to Larry La Frenchi, a local taxidermist. Neither La Frenchi nor a Wyoming state game agent called in as a consultant were positive as to what the animal was, so the carcass was frozen, then delivered first to the Billings, Montana (about 100 miles to the north), office of the U.S. Fish and Wildlife Service and from there to the F and W's wildlife research lab in Fort Collins, Colorado. There it was identified as a black-footed ferret, the first irrefutably seen since Granny passed away at Patuxent. The discovery was a very hopeful one for everyone concerned with the species, and the manner in which it was made a satisfaction to those who believe that a healthy respect for serendipity is the best antidote for hubris.

During the remainder of the fall, endangered species agents and a private consulting biologist, Dr. Tim Clark of Jackson Hole, Wyoming, sighted—or found convincing evidence of—about twenty ferrets in the vicinity of Meeteetse. (As of January 1983, it is estimated that there may be a population of fifty animals living in the area.) All of the animals were found in or near a 7,000-acre prairie-dog town, most of which was located on Pitch Fork Ranch, a 120,000-acre spread owned by Jack Turnell, whose family has been working the property since the 1870s.

With subjects once again available as grist, the federal ferret machine commenced grinding, the first motions being essentially political ones. In keeping with the strong states'-rights convictions of Secretary of Interior James Watt, the animals were effectively defederalized a few months after their discovery. Wyoming was encouraged to create a ferret advisory committee (on which there is one federal representative) and management of the creatures found near Meeteetse—or subsequently elsewhere in the state—was transferred from the U.S. Fish and Wildlife Service to the state body.

The head of the Wyoming group and thus the formal public guardian of all the now known living ferrets is Dr. Dale Strickland, the state's supervisor of Biological Services. Strickland is a very well regarded professional who shares the opinion of a growing number of younger zoologists that there has been excessive emphasis on hard "hands-on" research, especially in regard to endangered species. As the man now in charge, Strickland says his first responsibility is to protect the known ferret population from—among other things—overly zealous researchers; that he is dubious about the value of animals being picked up for clinical examination, radio telemetry studies and particularly for captive breeding programs. In his opinion the most immediate research problem is to try to determine what environmental elements have enabled the Meeteetse animals to survive and, relatively speaking, flourish. "If we can identify the significant factors which make this good ferret habitat, we are going to know a lot more than we do now about their natural history, how to protect them and where to look for other populations."

Jack Turnell, the owner of Pitch Fork Ranch, was also named to the Wyoming ferret committee. Since considerable numbers of the animals make use of the land he owns or leases, Turnell's cooperation is essential so far as their management and study goes. Like many rural westerners, Turnell is biased against the government—i.e., the federal one—and its agents, many of whom he feels "don't have much common sense." Also he does not have a good opinion of outside environmental groups and is not inclined to let them run around his place looking at it and giving gratuitous advice on what he should do about ferrets. However, he thinks very well of Strickland and is cooperating by providing an informal management history of the ranch which may give some clues as to why it is a ferret sanctuary. "We've just been running cattle here for years," he says, "and I guess whatever we have been doing has not bothered them much."

Turnell has a strong interest in the natural history of Pitch Fork, on which during the latter part of the 19th century the last free-roaming western buffalo survived. Not until after Shep made his unexpected find did Turnell, or anyone else, know there were ferrets in the area, but he is now very pleased that they are on his place. "Now that we know about them, we'll think about those animals when we are doing other things. They don't make any problems for us and we'll go on trying not to make any for them."

As compared to the previous operations of the endangered-species program, the present arrangements for preserving the Meeteetse ferrets is unconventional, more the responsibility of a few individuals than of the national bureaucracy. Setting aside general questions of precedence and ideology, this care-by-neighbors approach seems like a good one for these particular animals. The intentions of Strickland, Turnell, and their associates are fully as honorable as those of the federal agents previously in charge. The new guardians may err, as former ones have, but for the same reasons: because they don't know enough to avoid mistakes, not because they don't care enough about doing the right thing. At the moment the ferrets appear to be in a better situation than we have ever known them to be, but so far as anybody can tell, these are still creatures balanced precariously on the brink of extinction. If something goes wrong, it seems more fitting for both of the species principally involved that it happen on Pitch Fork Ranch rather than in a distant laboratory.

Turn of a Century

December 26 — The Seven Mountains

The Juniata River is to the south, the Susquehanna and West Branch of the Susquehanna to the east and north. The area is known locally as the Seven Mountains for the seven Appalachian ridges — Tuscarora, Shade, Jacks, Tussey, Nittany, White Deer and Bald Eagle — that run through it. The mountains rise in the west out of a great knot of Allegheny highlands and are separated by valleys through which flow small rivers and large streams that empty into the Susquehanna. Taken together, the highlands, ridges and valleys form a defiant fist of land 60 miles wide and twice as long.

    These are old mountains; there were towering peaks here when the land that was to become the Cascades, Sierras and Tetons lay under water. What is left of them is 2,000-foot nubs, skeletons of mountains. Their gnarled flanks are cut by mean, traplike ravines, littered with sharp ledges, pitted with sinks, oozing seeps and highland bogs. They are covered with a thick growth of oak, laurel and greenbrier that is as hard to move through as mesquite. The climate may not be the best or worst, but it is among the most unpredictable. In the summer the Seven Mountains are a jungle. A man trying to bushwhack up a ridge will sweat like a horse in the humid, stifling air. But snow and gales can come as early as October, come suddenly in a howling blizzard that drops the temperature 50° below freezing and piles hip-deep drifts in the hollows. Within a week a cold, driving rain may have converted the snow to fog, mist and slides of mud.

On a topographic map of the Seven Mountains there are extensive areas crossed only by trails, showing few if any signs of permanent human habitation. The empty places are designated as state forest or game land. This is such hard country that no one has been able to take much pleasure or profit from it.

The blank places on the map are honest ones. Many Indian tribes and nations hunted and fought through this country, but none were able or wanted to stay long enough to establish sovereignty over it. Europeans tried to break the mountains for more than two centuries. Yet it is still wild. It was here, in this hard fist of land, that a group of European peasants became American frontiersmen.

What happened on the Seven Mountains in the 18th century is seldom mentioned in popular histories. It has now become a folk myth, in part because events of that time and place tend to contradict popular history. For example, we have the notion that our forebears landed on the Atlantic Coast and immediately commenced their long but always triumphant progress to the Pacific. By virtue of their superior technology, ingenuity and grit, they overwhelmed the continent and its inhabitants and lived easily and well off the land. All of which is untrue. For better than a century, a third of the time white men have been here, they huddled on the coastal plains, unable or unwilling to leave the sea and their lifeline to Europe. They did not have the skills nor, frankly, the stomach to cope with the interior wilderness. They were pathetically dependent upon Europe for tools, weapons, clothing and even food, for their books, politics, religion, physical and psychic security. They did not try to find their way in the woods; instead, they hired or blackmailed Indians into guiding and caring for them. For their part the Indians apparently distrusted the Atlantic colonists because of their tactics and inclinations, but they were not in awe of them as men. For the best part of a century and a half the Huron, Shawnee, Delaware, Cherokee and the Iroquois Confederation, assisted by a few French advisers, rather contemptuously kept the more numerous Atlantic colonists pinned to their harbors and penned up in their fortified towns.

One difficulty was that the first emigrant boats were overloaded with gentry or would-be gentry who because of their pretensions and inexperience were too soft and squeamish for hand-to-hand wrestling with the wilderness. There was an oversupply of second sons, failed royalists, bankrupt shopkeepers, essayists, poets and a great excess of divines. In short, far too many chiefs and, so to speak, far too few Indians. White Indians, or at least those who had the makings of white Indians—Scottish, Irish and German peasants—did not begin to arrive until early in the 18th century with the second wave of immigrants, second class. The majority of these foreigners headed for Pennsylvania. There in the colony and City of Brotherly Love they were welcomed coldly by the local nabobs. "Bold and indigent strangers," said a Pennsylvania official of these scraggly newcomers. At the time bold meant uncouth and indigent meant immoral. "White savages," sniffed a young Ben Franklin.

In general the newcomers had the choice of living on the coast and remaining what they had always been—clients, tenants, servants of the gentry—or moving west beyond the reach of surveyors, lawyers and bankers. Many of them opted for the wilderness. In the second quarter of the 18th century they arrived on what was then called the Middle Border, the valley of the Susquehanna, in which stood the Seven Mountains. On this border, against the fist, they beat themselves and were beaten bloody for the rest of the century. In those early years they were scalped, raped, burned and starved; they died of fever, gangrene and exposure; they went mad from pain, murdered each other, became alcoholics and suicides. Yet because they were desperate for land and independence they stayed and learned to do what they had to do: how and why to take a scalp, to follow a deer trail, to kill deer, to make and wear buckskin, to jerk venison, to travel a week on a pocketful of jerky and corn, to use a double-bitted ax, to pry out stumps, to split logs. Among other things, because they had to have them, they invented what in later times and more romantic circumstances were known as the Kentucky long rifle, the Conestoga wagon and the bowie knife. They trained in this hard country, and utilized all they had learned there to move on, taking the whole continent in another 75 years.

Not only were peculiar tools and skills developed on the Middle Border but also a set of uniquely American attitudes: *The only good Indian is a dead Indian. Root hog or die. Fish or cut bait. That which is not useful is vicious.* The frontier tools and tricks have long since become obsolete, but the ideas are still in everyday use. If one were looking for the source from which still flows the mainstream of American culture and character, he would be well advised to leave behind the coastal athenaeums and boxwood mazes, where Europe petered out, and search among the Seven Mountains, where America began.

December 27—On the Buffalo Path

There is often both a nostalgic and smug, self-serving tone to place-names along the coast: Plymouth, Providence, New Jersey, Baltimore, Jamestown, Georgetown, Virginia, Carolina. From the Middle Border westward, names tend to be more contemporaneous and descriptive— Hungry Mother Mountain, Horse Thief Basin, Dead Indian Springs, Poison Spider, Hangtown; even such commonplaces as Fishing Creek, Middle Valley, Sugar Grove constitute a kind of spontaneous, topographic journalism. Read in this way there is a recurring theme to be found in the maps of the Seven Mountains. There are at least three ridges called Buffalo Mountain, a Buffalo Gap, a Buffalo Flats, a Bull Hollow. Lewisburg, a principal town in the area (the home of the Bucknell University Bisons), sits at the mouth of Buffalo Creek, which flows through Buffalo Valley in which there is the hamlet of Buffalo Cross Roads and a Buffalo Church. The names recall a largely forgotten fact, that until 200 years ago, and no one knows how long before that, the Seven Mountains was a pivotal area for herds of wood bison.

The Eastern animal was larger, darker and probably less numerous than the better-known plains buffalo, of which at one time there may have been 60 million moving together in great seas of flesh. Nevertheless, the wood bison were by no means rare. There may have been half a million animals in the Eastern herd that ranged from the Gulf Coast to Canada. The wood buffalo were migratory, moving north and south along the flanks of the Appalachians as the seasons changed. The bulk of the herd, which wintered in Georgia, Alabama and on the Gulf plain, would start north in the late winter, and some of the animals would continue until they reached the Great Lakes (thus Buffalo, New York). The Seven Mountains sat astride the principal migration route and also served as a major dispersal area. When the herd reached this point in the spring many small groups, called families by the Indians and later the frontiersmen, left the march, turned westward up the valleys and sought out small sheltered upland meadows where they foraged and calved during the spring and summer. These families numbered several hundred head of cows, immature animals of both sexes and always a few buffalo steers who had been castrated by wolves that hung on the flanks of the migrating herds. A big, experienced bull invariably led the families.

In the fall, when the migration was reversed, the Seven Mountains was a rendezvous. Trickles of buffalo would begin to flow east and south out of the mountains toward the Susquehanna Valley, where they would join other families and form the migratory river. It was said that in the fall the mountains rang with buffalo music, that the bull leaders would stand on the ridges, bellowing and, by inference, listening for the bellows of their distant colleagues. It was supposed that in this way the bulls were informed of each other's presence and progress, and would adjust their pace to meet at the Seven Mountains and without delay continue from there southward.

Being creatures of habit who generation after generation followed the same routes, the wood bison stamped out a series of broad trails through the Appalachians that were afterward used by all manner of other traveling creatures. Most of these trails are no longer recognizable as such. Some have washed away, some have been overgrown or obliterated by rockslides and floods. Some are modern roadbeds. (The lead bulls apparently had a keen instinct for contour.) However, here and there, especially in remote places such as the Seven Mountains, disconnected bits and pieces of the old trails remain.

What seems to be a surviving buffalo path crosses Nittany Ridge in one of the gaps of Seven Notch Mountain, wanders across tableland through a place called Buffalo Flats, intersects in a hemlock forest the headwaters of Buffalo Creek, follows it through a narrow mossy gorge called Buffalo Gap, down into the Buffalo Valley. The buffalo path is now infrequently used and is impassable for vehicles and horsemen. It is not even a good place for pleasure hiking. Laurel has encroached on the path and erosion has gulched across it.

Though in the valley it is warm, muddy, almost balmy, in the mountains a thin layer of ice, like grease on an old skillet, covers the buffalo path. Like so many things on the Seven Mountains there is an in-between quality to the sheath of ice that makes it difficult to move upon. It is not thick enough to hold crampons, yet too slippery to hold boots. In places the path is sunken, ditchlike. There are sizable ledges, flat shields of rock on the mountain, bare spots on which nothing grows but lichens. A man wanting to climb Seven Notch Mountain would have laid a trail more or less straight between these rocks. But the buffalo were in no hurry, and ate as they traveled. Therefore the buffalo path snakes around, often circling the rocks because at the edge of these balds in the sun there was more plentiful and succulent forage.

There is a boy—actually a young man—along on the journey. He has foregone holiday socializing, beer and girls to come out onto the slippery Seven Mountains. It seems that for his sake something should be said to emphasize this faint, overgrown path.

"You know, this is what they call a primary historical record. There were no books written about it that have lasted as well as this path the buffalo made."

"Or tell you as much about buffalo."

December 28—Above Buffalo Gap

There once was a saying that when the first redbuds bloomed on Bald Eagle Mountain you could look for the herds of wild cattle moving north and west, and that they returned in the fall when the persimmons were ripe. The Middle Border settlers went out to look for them with guns and knives, killed them for the meat and hides and to keep them away from their clearings and crops. There were men who could brag of having killed 2,000 buffalo, which meant that at least sometimes the animals were killed for fun and the tongue. Nobody could make use of the meat and hides of 2,000 buffalo, and there was little trade since there was no dependable way of shipping them east.

In consequence the herds rapidly became smaller and their migration pattern was broken. By the 1780s the craftier or perhaps more timid lead bulls refused to run the gauntlet of guns. They no longer made the semiannual rendezvous in the dangerous valleys. With their families they remained high up on the mountains and kept to their summer ranges the year around. They had no other choice, but it was a doomed response. There was not enough forage in the highland pastures to support continuous browsing, and the animals probably starved by the hundreds.

What wolves and panthers were left, themselves cut off from their former range and prey, must have attacked the declining buffalo with increasing boldness and desperation. Finally, while the retreat into the highlands may have made the work of the valley hunters harder, it did not deter them. They would locate a buffalo family on the ridge, surround it and kill as many as they desired, then pack the meat and hides they wanted down to the settlement.

By the winter of 1799 only one herd of buffalo remained on the Seven Mountains or, as it later developed, in all of Pennsylvania and very likely in the entire northeast quadrant of the continent. This family ranged the ridges on both sides of Buffalo Valley and was led by a bull who had been named Old Logan after the Iroquois war chief. Logan the Iroquois was described as the "most martial of all Indians" and "a man of superior talents but of deep melancholy to whom life had become a torment." He is best remembered as the author of *Logan's Lament,* a dirge that was publicized by Thomas Jefferson. *Logan's Lament* was spoken over the bodies of 13 of his family who had been murdered by Middle Borderers. It went, in part, "There runs not a drop of my blood in the veins of any living creature." Having mourned, Logan went to war, and is reported to have taken precisely 13 white scalps. He was killed in 1780, either bushwhacked by whites or by a tribesman acting as their agent.

Old Logan, the buffalo, was said to have been a coal-black bull of exceptional size, wariness and ferocity. Sometime in the late fall of 1799 someone had come across his herd deep in the mountains and counted them. Thus it was known that at the last the bull led a family of 345 animals.

In the flats above Buffalo Gap it is the kind of day hereabouts called iron cold. It is a descriptive phrase. The heavy low clouds are gunmetal gray, and even at noon there is no warmth in the sky, much less on the plateau itself. Buffalo Creek flows through bands of ice, and both the ice and water are metallic. Even sphagnum moss does not rustle or squish underfoot but cracks. Hoarfrost breaks the ground like crystalline fungus. The hemlocks stand stiff and rigid, and their limbs snap in the wind. On this day there are few living things to be seen—two chickadees and a cruising crow. It seems that such a place in such weather could not support much more, but in fact Old Logan and his family might have lasted out this kind of an open winter as they had others, eating bark, moss, scrub bushes and the precious few bunches of frozen bog grass. But the buffalo's luck was bad. The winter of 1799 was a terrible one, even for the Seven Mountains. The blizzards came after Thanksgiving, and there was no thaw. By Christmas the buffalo family must have been starving or so nearly so that their hunger overcame their fear of the valley. On Christmas Day or thereabouts Old Logan led the herd down off the drifted flats.

December 29—Buffalo Field

Half a mile to the west of the crossroads at Port Ann, in Middle Creek Valley, there is a knoll almost under the wall of Jacks Mountain. Long ago this was called Buffalo Field, but now it is spoken of as "the place where the distillery used to be." The descendants of its first proprietor live in his farmhouse. "They stored the kegs in here," says his great-granddaughter. "I suppose this might be regarded as a historic place, but the fact is that it was a gathering place for drunks. They came for the free whiskey my grandfather and his father passed out. Then they passed out."

At least three-quarters of a century before the distillery was founded, a man named Samuel McClellan built a cabin on this knoll under Jacks Mountain. "There are still some McClellans in the valley," she says, "but I didn't know they had lived here. However, now that it's mentioned, it seems I heard, a long time ago, that story about the buffalo. Or maybe I just imagined it."

So far as recorded, or even folk history is concerned, there were only three important days in Samuel McClellan's life—the last three of the 18th century. However, because of those three days is is possible to guess other things about McClellan. He was probably then a youngish man, since he had a young wife and three children, all under five years old. It is likely that he was poor, as all the McClellans lived in an insubstantial one-room cabin that was not yet fenced and did not have outbuildings. He may have been a newcomer, at least to Middle Creek, but he had a good Middle Border name and had picked up at least one of the area's habits. It was not snowing on the morning of December 29, 1799, apparently a rarity for that winter, so McClellan had taken advantage of the break in the weather to go down to the creek to cut wood. When he went, he took his gun.

With the wisdom of hindsight, we can now see that it was unfortunate that McClellan carried a gun that morning. He had been working only a short while when Old Logan, followed by his starving family, came snorting down the frozen creek bed, looking for food and survival. McClellan promptly killed four cows. Had things gone otherwise, he probably could have made good use of the meat, but his immediate intentions were most certainly defensive: to turn the herd away from his cabin and those of his neighbors. However, the 341 remaining buffalo stampeded down the creek until they came, with McClellan laboring along behind, to the establishment of Martin Bergstresser, a more substantial place than that of McClellan. There the buffalo, crazed with hunger, broke through a stump fence and lumbered straight to a pile of hay, Bergstresser's entire store of winter feed for his own stock. They demolished the mow in a matter of minutes, and in the process flattened a fence, a springhouse and stomped to death, so it was remembered, 6 cows, 4 calves and 35 head of sheep.

Even if the story was somewhat exaggerated in the retelling, this one incident should make it clear to any but the most incurable romantics why there have been no wild buffalo for nearly a century. Bloodlust, greed, meat and hides were secondary factors. Just one of these beasts could shred a fence, or knock over a gas pump, for that matter. And 341 of them could ruin a man or, in the right circumstances, a settlement. As for 60 million buffalo, the capricious energy locked in the great herds was that of an avalanche. Given what we are, we could no more share the land with them than with a wildfire. It is sometimes argued that it is a pity we became what we are, that the land would be gentler and prettier if the few of us who could live in that way were a nomadic, hunting and pastoral people. That may be true, but it is beside the point, since long before 1799 we chose otherwise. Once the decision was made, the buffalo, among other things incompatible with our ambition, was doomed.

These facts were underscored by Old Logan's family a few minutes after they had demolished Martin Bergstresser's barnyard. Bergstresser, his 18-year-old daughter Katie and McClellan killed four more of the animals, but the herd stayed until they had finished the hay. Then, pursued by the two men, the girl and a pack of yapping dogs, they fled back upstream and shortly came to the clearing around McClellan's cabin. There was no hay there, so perhaps it was sheer confusion that made the buffalo halt and stand in a milling, pawing circle in the cabin yard. Old Logan stood facing the cabin door. From inside, above the sound of the buffalo, could be heard the screams of McClellan's wife and children. Having run out of shot, McClellan rushed through the herd and, in an effort to turn the bull, attacked Old Logan with his bear knife. Old Logan charged, not the man but the cabin, crashing through the flimsy door. He was followed inside by members of his family until, as Henry Shoemaker wrote, "They were jammed into the cabin as tightly as wooden animals in a toy Noah's Ark."

But then the commotion had drawn several other neighbors, and together the men began to tear down the cabin walls. When they had opened a side, the buffalo ran out "like giant black bees from a hive." Inside, the men found the bodies of McClellan's wife and babies trampled into the earthen floor. It was said that nothing larger than a handspike remained of the interior furnishings.

McClellan's lament is not remembered. Perhaps he never made one. However, one reaction was as predictable as that of Logan the Iroquois.

Above everything else, Samuel McClellan, standing by the wreck of his cabin and his life, must have thought of vengeance, and perhaps in these first moments he was mercifully numbed by this desire. McClellan took a loaded gun from one of the neighbors, and as Old Logan emerged, he shot and killed the big bull. Shortly thereafter McClellan and Bergstresser, on borrowed horses, rode off, one up, the other down Middle Creek Valley to raise help. Others surely would have made the ride, but perhaps the greatest kindness they could show McClellan was to let him ride off alone, beating a horse through the snow.

December 30—Jacks Mountain

The gray clouds above Buffalo Flats have fulfilled their promise. A cold steady drizzle begins to fall during the night. It takes an act of will even to get up on such a day, a constant repetition of the act to stay on Jacks Mountain. The hemlocks, pines and bare oak all are heavy and dripping. The trails are beds of mud, streams of slush and icy water. There is not one dry, warm, cheerful place or moment on the ridge. It is weather that defeats good gear and good intentions. Neither the body nor the mind can escape or ignore it for long.

It must have been five or six degrees colder in this same place on this same day in 1799 because snow was falling then at the rate of an inch or so an hour. It was no worse for ordinary living or travel than the rain. But, it was worse for the special business—pursuing the buffalo—that occupied the Middle Creek settlers. That morning 50 men gathered at Martin Bergstresser's ruined farm. The names of many of them were recorded: Ott, Snyder, Sourkill, Young, Doran, Everhart, Fryer, Jarrett, Middleswarth, Benfer, Miller, two Fishers, three Swinefords. They are the names still found in valley graveyards and on valley mailboxes.

They had a hard hunt ahead of them. The new snow was deep enough to have covered the tracks of the herd. They had to go on foot, since— even if they had had them—horses would have been useless in such weather. Finally, though again it may outrage historical fancy, they were not as well prepared for mountaineering as even the casual, occasional weekend hiker of today. They would have been wearing heavy deerskin coats, buffalo robes, heavy stiff boots, perhaps moccasins that soaked up ice water like a sponge. They would have carried heavy axes, knives and muzzle-loaders. Since they intended to stay out until they found the herd, each man would have carried provisions—a sack of corn dodgers, some grease, maybe a little piece of meat. Even on such a hunt it would have been surprising if at least a few did not calculate whether the comfort of a stone jug was worth its weight.

How they hunted, whether they split into smaller groups to cover more ground or were confident enough to guess where the buffalo would go to stay together, is not remembered. All that is known about that day is that they did not find their quarry, and that they slept the first night in the snow on the mountain.

December 31—The Big Sink

Nobody remembers who first said, "If you don't like the weather here today, wait until tomorrow," or where he lived, but if he were not a Seven Mountains man, he should have been. The storm has passed quickly, and just as quickly the temperature has dropped close to the zero mark and a stinging, boring northwest wind blows. It is likely that the last day of the 18th century was an identical one. It must have become bitterly cold during the night because in the morning, when the hunters started out again, the drifts were glazed over with a layer of ice thick enough to bear the weight of a man and, as it turned out, thick enough to freeze a buffalo in its tracks.

The avengers found the herd, presumably about midday, in a place that was then called the Big Sink. The name has disappeared from local usage and maps, but if it were not what is now called Bull Hollow (the name as well as the topography is suggestive), it was a place very much like it. Bull Hollow, a narrow, swampy, hemlock-choked ravine, is less than half a mile long, hollowed out of the ridgetop at the confluence of Jacks and Thick Mountain. A series of small seeps and springs rises to the west and forms a small creek that flows through the hollow. The walls of the hollow climb steeply 200 feet or so. From the ridge above, even on a bright cold day, the bottom of this mini-gorge is a dank, gloomy Transylvanian-looking place.

Despite its corrallike features, the Big Sink was, given buffalo experience and instincts, a logical last refuge. The weather having been bad for so long, some of the herd may not have eaten in nearly a week except for a few mouthfuls of Martin Bergstresser's hay. Also, with the mysterious weather sense many animals have, they may have felt the coming blizzard and approaching cold of the next day. Finally, they probably were terrified by the men, guns, dogs and the loss of Old Logan and the eight other animals. Under the circumstances, the small gorge was perhaps as attractive a place as they could have reached in two days. It was wild and isolated and the walls would have given some protection from the wind. A few winter greens and succulents might still have been growing around the seeps and could have been pawed out of the mud.

For whatever reasons, sometime during the blizzard of the 30th the herd filed into the hollow and remained there, dumbly enduring as the storm passed and the ice formed on and around them. When the hunters came to the ridgetop and looked down into the sink they saw the remaining animals locked in place by the crusted drifts. The men slid down the sides of the hollow. At first they killed the buffalo with guns, but when the extent of the great beasts' helplessness became apparent, they found it easier and less expensive in terms of powder and shot—perhaps even more satisfying—to come at them over the ice, hacking them with bear knives. They cut out the animals' tongues and stuffed them into the great pockets of the deerskin coats. The job was not finished until dusk. The last wood bison herd in Pennsylvania, the last anyone was to see in the Northeast, was still on its feet, held upright by the ice. However, the buffalo were all dead or dying, their broken jaws hanging agape, their throats tongueless. It was said, and certainly must have been true since the weather had not moderated, that the ice in the bottom of the sink "resembled a sheet of crimson glass."

When they were finished the men climbed back to the ridgetop. There they pulled together a large pile of dead wood and lit it as a signal to those waiting in the valley below that vengeance had been had, that the buffalo were no more. Later that night, perhaps after they had roasted some of the buffalo tongues, the party marched down into the valley, it is remembered, singing hymns. There cannot have been another New Year's Eve procession like it—50 blood-soaked men, cold with winter and grief but inevitably hot from the excitement of slaughter and self-righteousness, singing as they walked through the night down a frozen mountain into a new century. Yet, despite the portentousness of it all, it seems like a mistake to look for or force a moral on the history of Old Logan, Samuel McClellan, Logan and their families. True tragedies are not morality plays. They are always stories of necessity.

Addendum

The extraordinary events leading up to the killing of the last herd of wood bison on the last day of the 18th century became a tale, to be told and retold on the Seven Mountains throughout the 19th century. One day in New Berlin, in 1903, Flavel Bergstresser, the great-great-grandson of Martin Bergstresser, told it to Henry W. Shoemaker, a noted Pennsylvania folklorist and author. In three essays, *A Pennsylvania Bison Hunt, Extinct Animals of Pennsylvania* and *More*

Pennsylvania Mountain Stories, Shoemaker recalled portions of this conversation. There are certain omissions and some discrepancies in the narrative, which is not surprising since when he spoke to Shoemaker, Flavel Bergstresser was a man of nearly 80 years, telling of something he had been told had happened more than a century before. Drawing upon other records and circumstantial evidence, and by making certain deductions, some of these gaps have been filled, creatively but not capriciously. But my main debt is to Henry Shoemaker, not only for preserving the story, but for having in a sense been a winter companion and guide on the journey to the Seven Mountains, along the Buffalo Path, in Buffalo Field and the Big Sink.

Old Fish

Sport fishermen, riparian ecologists, and authors of guidebooks invariably refer to the "prehistoric paddlefish," as if this were a common name like Dolly Varden trout or smallmouth bass. Technically it is not inaccurate, since all living creatures, ourselves included, are, in a sense, prehistoric. But it is inadequate—for the same reason it would be inadequate to describe Tutankhamen as "the late art collector."

The paddlefish is exceptional in that it has not changed much in three or four hundred million years. In fact, paddlefish may be the oldest big animals surviving in North America. *Polyodon spathula* (as the species is formally known) was swimming in the warm continental waters when today's coal deposits were still forests of tree ferns. Paddlefish evolved way before the perch and pike, gar and grayling, and our other true fishes. They were here fifty million years or so before the first dinosaurs. Compared with *P. spathula,* the natural history of mastodons, mockingbirds, men, and other recent experimental models is as today is to last year. Prehistoric indeed.

By what we call "historic" times, i.e., after the advent of European-style observers and record-keepers, paddlefish inhabited only the rivers and sloughs of the Mississippi Basin—from the larger tributaries of the western Appalachians to those of the eastern Rockies. (There is only one similar species, in the Yangtze River of China. Why these two closely related creatures should live so far from one another is a mystery.) In our major midland waterways the paddlefish did well. There are old records of two-hundred-pound, six-foot-long fish and rumors of much larger ones. These days a one-hundred-pound, five-foot paddlefish is considered large. A century ago, and for many hundreds of years before that, the paddlefish apparently was one of the most abundant species, and because of its size and numbers made up a considerable portion of the riparian biomass.

In comparison with other fish *Polyodon* are primitive in appearance and behavior, but given their longevity and success it is unseemly to be patronizing about the adaptations they have made. The most visible of these adaptations is the snout, which has given the creature its common names (spoonbill and spoonbill cat, among others). This snout is up to two feet long, a rigid protuberance that looks something like a narrow beaver's tail or a fat cake spatula. In times past it was believed the bill was used to scoop out deep, secure resting holes in the bottoms of muddy rivers. Current biological thinking is that the paddle is a kind of antenna in which there are sensory organs that enable the fish to detect and react to water currents, underwater obstacles, and the topography of riverbeds. How this apparatus functions is not clearly understood. However, the advantage of having some sort of nonvisual direction-finding device is obvious, since paddlefish operate for the most part in the deep waters of big, muddy rivers. Not surprisingly, their eyes are tiny, pea-sized organs.

Underneath the frontal paddle there is an impressive mouth which when opened encompasses an area equivalent to that of a gallon bucket. Generally it is open, since the paddlefish feeds by cruising along, sucking in water containing microscopic animals: daphnia, diatoms, rotifers, and others collectively known as zooplankton. The water, silt, vegetable matter, and other debris (occasionally including a minnow or fishhook) are filtered through an intricate, sieve-like arrangement of gill rakers which serves the paddlefish as baleen does the great whales. The zooplankton is retained and ingested.

For the first few months of life paddlefish have a full set of tiny teeth; however, by midsummer of their first year they have shed these baby teeth in favor of gill rakers. During the summer, young paddlefish— which in April are approximately the size of newly hatched tadpoles— grow rapidly and have been known to attain a length of twenty-eight inches by October. This impressive rate of development is not maintained, but no one is certain what the limits of growth for a paddlefish may be. Tagged individuals are known to have lived for at least thirty years, and the assumption is they can live much longer. Apparently they continue to grow as long as they live, and they feed almost constantly. Therefore biologists believe that the nineteenth-century stories about three- and four-hundred-pound individuals cannot be dismissed out of hand.

In addition to being toothless, mature paddlefish are scaleless, with a smooth, green hide as tough as heavy-duty vinyl. Underneath the skin is a thick layer of meat that is red because it is finely laced with blood vessels. This layer has a lung-like function and helps the animal to absorb oxygen in the deep, turgid waters it prefers. Also, in part because of this characteristic, paddlefish are strong and durable swimmers. They may travel five hundred miles or so to spawn and can move along at three miles per hour against a current running at the same rate. For short bursts, they may sprint along at ten miles per hour.

Underneath the red vasculated layer, the flesh of the paddlefish is white, firm, and, like that of sharks, boneless; the paddlefish is stiffened entirely by cartilage. The lack of bone, plus the fact that the paddlefish has a tail remarkably like that of a shark, led to speculation that it might be a primitive or degenerate freshwater shark. Now, however, it is believed that sharks and paddlefish simply made similar environmental adaptations but are not connected in any direct evolutionary way.

During their first year or so, paddlefish are vulnerable to predation by other carnivorous fish, birds, and occasionally mammals, but thereafter their size makes them inconvenient food for anything but parasites. Additionally, because of what they feed upon and how they feed, competition between them and other large river species is minimal. Paddlefish apparently slurped along through the big rivers, not bothering or being bothered by anybody from the late Devonian period until the nineteenth century. Then they met us, and the hard times began.

As was the case with the grizzly bear, rattlesnake, and Yellowstone geysers, the first Europeans who met paddlefish told a lot of frightful stretchers about them. Early frontier accounts of paddlefish had them overturning rowboats, swallowing small livestock and children, ramming and jamming the works of paddlewheel steamers. However, rivermen began taking a more clinical and exploitative look at the fish and discovered that in spring a big female might hold as much as twenty-five pounds of eggs similar enough to those of the scarce sturgeon to bring caviar prices. In consequence, by the 1890s two million pounds of paddlefish were being netted annually in the Mississippi, Ohio, and Missouri rivers. They were taken principally for the roe. The flesh, if marketed, was usually dried or smoked, since fresh paddlefish meat was then generally considered coarse and rank. The paddlefish trade was unregulated, and by 1907, when the first formal study of the species was published, the catch had declined drastically. Since then there never have been enough big paddlefish to be of much commercial importance.

In addition to overfishing, certain environmental alterations contributed to the paddlefish's decline. Among the most critical were the massive flood control, hydroelectric, and channelization projects that have changed the characters and courses of nearly all our big midland rivers. The rivers have been "improved" for people by straightening them, removing obstacles, squeezing them into narrower channels, and increasing the rates of runoff and flow. By way of illustration, the Missouri River, a notably unruly one, now occupies about half the surface area it did before the Army Corps of Engineers began "improving" it, and it flows almost twice as fast. Much of the riverbed that was eliminated was prime paddlefish habitat. Paddlefish thrive in deep, slow, meandering backwaters where the zooplankton is rich and where the lack of current enables them to feed efficiently — the type of habitat the Corps and other public agencies found most untidy. For example, islands create the kinds of pools in which paddlefish prosper. However, all but 18 of the 161 major islands in the Missouri have been eliminated. The situation is similar elsewhere.

Until recently no one knew the species was being affected. Commercial fishermen had lost interest, and sportsmen never had much interest in a fish that did not take a bait or hook. Therefore there was no public clamor about the status of the paddlefish and no pressure on wildlife managers or researchers to look into the matter.

The situation changed abruptly in the 1930s when Bagnell Dam, a 150-foot-tall hydroelectric facility, was built across the Osage River, a major tributary of the lower Missouri. Bagnell Dam produced much of the electricity used in the St. Louis area, and it created the Lake of the Ozarks, a 60,000-acre impoundment which has since become one of the notable recreation areas and tourist traps of the Midwest. The project also had a profound and unanticipated effect on paddlefish. The creation of Lake of the Ozarks set in motion a chain of ecological events which resulted in:

What appeared to be a paddlefish population explosion;

The emergence of the paddlefish as a popular game species and the consequent creation of a pro-paddlefish constituency;

Considerable investigation into the natural history of the fish;

The present possibility that the paddlefish, which has survived for a third of a billion years, may not make it into the twenty-first century.

It is now believed that the Osage River, just prior to the construction of Bagnell Dam, may have supported the largest, most vigorous population of paddlefish remaining in the Mississippi Basin. When Lake of the Ozarks was filled, many of these fish were permanently trapped behind the dam. The lake provided the fish with the kind of habitat they liked best, and they multiplied rapidly. They became very visible, especially in spring. Schools of paddlefish, larger than anybody had ever seen, congregated in the narrow upper reaches of Lake of the Ozarks. This same increase in paddlefish populations immediately following the creation of impoundments had been observed on other rivers, but generally it was a short-lived phenomenon. In most cases the fish would begin to decline after a few years and then disappear. However, in Missouri they remained and continued to do well.

Lake of the Ozarks was a boon to fishermen as well as paddlefish. In the mid-1930s it was discovered that paddlefish—which cannot be taken by conventional means—could be snagged by dragging big triple hooks through the channels and holes where large numbers of them gather in spring. Once hooked, a paddlefish provides a considerable amount of excitement. The fish's immediate reaction, according to veterans of the sport, is to make a strong run of fifty yards or so during which a big individual can pivot a johnboat and sometimes tow it. Thereafter the snagger, using 110-pound-test or strong line, may play the fish for half an hour or so until both parties are exhausted and the fish is boated.

Besides being a sport fish, the paddlefish is good eating. Formerly the fish had been held in low esteem because nobody had learned to clean them properly. The trick, Ozark fishermen found, was to remove the outer layer of red, vasculated meat and eat only the boneless white meat underneath. This meat resembles that of swordfish and is excellent by any standard.

Ever since 1960, between fifteen and twenty thousand snaggers have come to Lake of the Ozarks each spring, taking four thousand or so paddlefish a season. There is some snagging as far west as Montana, but Missouri remains the center of the sport because the Osage drainage system is the only place where there is a sizable paddlefish population. The paddlefish rapidly acquired a consumer lobby, and public agencies—specifically the Missouri Department of Conservation—are giving it more attention.

Twenty years ago, little was known about the reproductive cycle of the paddlefish. It was assumed that the springtime schooling in Lake of the Ozarks was associated with spawning, but there was no biological information about when, where, and how this took place. Even more puzzling, nobody had seen paddlefish fry or fingerlings. Back in the 1920s the mystery of where the baby paddlefish were had so intrigued Edward Phelps Allis of Allis-Chalmers that he offered a $1,000 reward to anyone who could find a paddlefish as small as two inches. The money was never claimed, and the reward was withdrawn. It was not until the spring of 1960 that Charles Purkett, a Missouri fisheries biologist who had been investigating tributary streams above Lake of the Ozarks, located the first spawning paddlefish. Purkett described his discovery in the May 1961 issue of the *Missouri Conservationist:*

On April 20, the seventh day after a nine-foot rise, and for several afternoons and nights following, fish were observed in what we later proved to be spawning activity over the gravel bars near the mouth of Weaubleau Creek. The only evidence of spawning was the appearance of a single fish at the surface where it would agitate its tail rapidly, then disappear after a few seconds. It appeared this was a female that had started near the bottom over a large gravel bar and made a spawning "rush" during which eggs were released. Presumably, accompanying males then released milt which fertilized the eggs. This happened every few minutes during the afternoons and evenings while the water level remained high. . . .

By the afternoon of April 24 the water level had fallen seven feet very rapidly, exposing the gravel bar where spawning had been observed. That afternoon I walked to the area where most of the activity had been and picked up a paddlefish egg almost immediately — and dropped it into the flowing water! Fortunately there were many more just downstream, and as I searched I also found hatching eggs and some which had hatched only a couple of hours before. . . . .

Gravel bars which are kept clean by flowing water and are deep enough for the thirty- to seventy-pound fish are preferred for spawning. Much of the gravel and rocks in these bars is small and is not silted over as it would be in the lake. This is the reason these paddlefish make the long trip up the lake and river. . . .

The eggs, when released from the female, flow freely and do not stick to anything. Immediately after fertilization a very adhesive coating forms which causes the egg to stick to the first object it touches. Most eggs adhere singly to a pebble or rock so strongly that some eggs were broken in trying to remove them. Some were coated with silt before they became attached, and some attached to floating sticks and were washed downstream. We dredged up quite a number of these in places where they probably wouldn't have hatched and survived. The most favorable location for hatching obviously was the clean-swept gravel bar. Here they were not covered with silt and aeration was good.

Subsequent studies indicated that paddlefish need a very specific combination of conditions for spawning: clean gravel bars that are briefly inundated by spring floods, floods that provide at least a five-foot rise in the stream level, and a water temperature of about 60 degrees. This combination occurs for only a few days in the spring, but not every spring. In years when conditions are not suitable, female paddlefish resorb their eggs, retreat downstream, and do not spawn.

This information made it rather obvious why paddlefish had done very well in Lake of the Ozarks but badly in most other areas of the Mississippi Basin. The waters backed up behind Bagnell Dam did not extend far enough upstream to silt over the gravel spawning bars. But in the other rivers these breeding grounds, as well as feeding areas, had been eliminated or degraded by impoundments and channelization. Paddlefish populations, sharply reduced at the turn of the century by overfishing, could not recover and were growing steadily smaller. The lack of early records makes accurate comparisons impossible, but Missouri biologists speculate that the Mississippi drainage system may now support only 10-20 percent as many paddlefish as it once did. In many rivers, particularly in the eastern half of the basin, the fish is now extinct.

Unfortunately, the paddlefish sanctuary in Lake of the Ozarks proved temporary. In 1954 the Corps of Engineers proposed the Harry S. Truman Dam, to be built on the Osage River some fifty miles above Bagnell Dam. The resulting impoundment would effectively double the size of the Lake of the Ozarks complex and flood out the gravel bars on Weaubleau Creek and those on all the other tributaries where paddlefish were known to spawn. After Purkett and other researchers documented this, the fate of the paddlefish became a considerable issue in the controversy about whether the dam should be built. However, in 1973 a federal court gave final approval for the project, and in late 1977 Truman Dam was finished and Lake Truman filled.

During the past three springs paddlefish have congregated below the new Truman Dam, vainly seeking to reach their traditional spawning beds above it. A snagging season has been permitted both below and above the dam on the grounds that these fish are effectively sterile, since they now have no place to reproduce.

The environmental changes caused by Truman Dam scarcely came as a surprise, since the project had been under argument and construction for nearly a quarter of a century. In the 1960s the Missouri Department of Conservation began considering what it would do after the impoundment destroyed the paddlefish spawning beds, and opted for trying to rear the fish artificially. After considerable experimentation, this is now being done at the Blind Pony Hatchery near Sweet Springs in the western part of the state. The hatchery biologist, Jerry Hamilton, is regarded as something of a genius for having found out how to hatch and raise paddlefish, something nobody else had done previously and which thus far has not been done elsewhere with much success.

In spring, as paddlefish begin their spawning run in Lake of the Ozarks, Hamilton and his associates net thirty or so gravid females and a dozen mature males. The females weigh between thirty-five and sixty pounds, and each may contain half a million eggs. At the same time, Hamilton and his crew make arrangements to take a quantity of pituitary glands from fish caught by snaggers. These glands are immediately frozen and stored.

The captured fish are held in small ponds at Blind Pony Hatchery and observed closely until Hamilton judges the females are ready to lay their eggs. Then each female is moved into a laboratory holding tank and injected with pituitary material extracted from the frozen glands. Without this injection of hormone material the sows will not release eggs.

Paddlefish eggs, unlike those of nearly all other freshwater species, are not contained in a single sac and deposited at one time. Rather, the roe is held within the ovaries in a series of flat "leaves," which are released individually over a period of twenty-four hours or longer. In the hatchery, females that have been induced to begin laying are tended constantly, and as a leaf is released, the eggs are immediately fertilized with milt extracted from one of the captive males. Eggs are then held, under constant temperature, in large test tubes. They hatch in about seven days. The fry are transferred to outdoor rearing ponds, rich in zooplankton, and kept there until early fall.

Despite the great quantity of eggs produced by each female, the survival rate of fry, at least with captive fish, is not high. Thus far, only about 150,000–200,000 fingerlings are being raised at the hatchery each season. However, Hamilton believes this figure could be increased to half a million. A small number of paddlefish have been hatched in South Dakota, but the Blind Pony laboratory is the only place where they are produced regularly in quantities. Fry and fingerlings have been shipped from there to other states and also to the Soviet Union, where it is hoped dwindling sturgeon populations can be augmented—and the caviar industry supported—by the introduction of North American paddlefish.

In 1972 the Missouri Department of Conservation began stocking paddlefish in Table Rock Reservoir, a large impoundment along the Missouri–Arkansas state line that was selected as a controlled testing site because it previously had no paddlefish. The habitat there proved suitable, and Table Rock now has an established though presumably nonbreeding population. In 1978 the first ten- to thirteen-inch fingerlings were released in Lake Truman, and during the next several years Lake of the Ozarks will be stocked.

Kim Graham and Tom Russell work full-time as paddlefish field biologists in Missouri. Graham is reasonably confident that a respectable number of the fish, enough to support a sport fishing season, can be maintained in the Osage River impoundments, Table Rock, and perhaps other new areas. However, he feels that there is little chance these can be anything other than stocked fish, hatched at Blind Pony. "There may be a few small gravel beds in Missouri where there is natural reproduction," says Graham, who has searched persistently for such areas, "but we don't know where they are and haven't seen any that looked very promising. As far as we know, this is the situation throughout the Mississippi Basin. Spawning beds are scarce, and prospects are that there are going to be fewer rather than more of them in the future."

The sterile, brightly lit laboratory building at Blind Pony Hatchery is perhaps the best paddlefish spawning ground left in this country. In it, in a steel tank, lies a gravid, fifty-pound sow that has been captured, transported, manipulated, probed, and stimulated by attending biologists and technicians. If their techniques continue to be effective; if the thermostats, water filters, circulators, and air conditioners are properly maintained; if secondary support systems, telephones, combustion engines, highways, petroleum refineries, and electrical generating plants continue to function as planned; if legislative appropriations committees, bureau supervisors, taxpayers, and voters remain supportive; if inflation or depression, and political, social, or international crises do not disrupt everything, this paddlefish will give up her eggs, and some of her daughters will do the same in the same way. The eggs will hatch, and the line of the species, which has survived the settling of continents, the lifting of mountains, the ages of ice, coal, and reptiles, may continue in the late Engineering Age.

That we have figured out how to do these things to and for such an ancient creature is a credit to our intelligence. That we want to do them speaks well for our compassion—or at least says something about how well we regard sport. That we need to do these things—that the paddlefish, her attendants, and the rest of us have met for these purposes at Blind Pony Hatchery—is unspeakably sad.

Part II: In God's Countries

Even when I have not had the wit or industry to avoid using it, the adjective natural — as in natural areas, natural food, natural law — has exasperated me because it implies the existence of unnatural phenomena, which I do not think there are. Especially, unnatural has become a code word for humans and their activity. The idea that we are unnatural is a long-standing philosophical and theological one which rises from two superficially disparate sources. The first is that we have been placed by some undescribable being or process above the rest of the known world and have been given dominion over it. The second is that we are below Nature, less pure and good than, the corrupters of, it. In both conceits we are the only species which is estranged from the terrestrial environment — the only home we are indisputably known to have. Accordingly, we either do not have to obey the house rules because of our superiority or cannot because of our inferiority. This has alternatively and sometimes simultaneously produced a lot of hubris, guilt, and loneliness which I think is undesirable and unnecessary. There is no evidence that we do not belong here. In regard to the rest of the world we can certainly do damn fool things that appear — at least in the short term, which is all we can deal with — to be in nobody's best interest, including our own. However, I am not convinced we can commit unnatural acts. In this sense the following are down-home stories about some of the sections of our natural manse, their and our attributes.

The Rites of Autumn

The Armed Invasion

On the Monday following Thanksgiving in the state of Pennsylvania a good many commercial, governmental and educational enterprises are shut down or critically understaffed because most of the males over 12 years old have grabbed their guns and left. Some of the men are in convoys of covered trucks and miscellaneous motorized carriers that clog the highways and back roads; hundreds of thousands of others, more than can be mustered by the U.S. Army, are already skirmishing through the countryside. They blaze away more or less at will. The sound of small-arms fire is incessant. At dusk they fall back to rude bivouacs and commence the body count.

Appearances aside, the Commonwealth is not in the grip of foreign invasion or civil insurrection, but only of Opening Day, an event of such magnitude that in the Quaker State nobody asks, The opening day of what? Phillies? Steelers? Nittany Lions? Opening Day is the first day of the buck-hunting season. It may be the wildest and woolliest — and it is almost certainly the largest — participatory recreational event extant. Pennsylvania has about a million and a quarter hunters pursuing a herd of 700,000 deer, and the result is a kill of some 150,000 in a good year, with about 60 percent of the animals usually being scragged on the frantic Opening Day.

Because white-tail deer are commonly distributed from the Philadelphia suburbs to the strip-mine-scarred hills above the Ohio River, the Opening Day uproar is general. However, it rises to a crescendo in

Potter County, a mountainous, semiwilderness district in the extreme northern section of the state. Every year some 50,000 sports go deer hunting in Potter, and among them they do in 7,500 animals, year in and out the largest harvest of any county in the state. As this kill total implies, Potter has a lot of deer; in fact, the common wisdom is that the 16,395 permanent human residents of the county are outnumbered by at least two to one by the white-tails.

In pragmatic terms, however, Potter leaves something to be desired for both deer hunters and deer. As to the former, some 500,000 acres of the county's gnarled mountains are covered by dense stands of hardwoods and evergreens, a region known locally as the Black Forest. There are a lot of places for deer to hide in this terrain, and actually shooting one is complicated by the fact that 50,000 hunters all have this as a common purpose. Furthermore, once a Potter deer is taken, it is not likely, by conventional sporting standards, to be a very impressive animal. Popular myths aside, heavy forest doesn't make particularly good deer country, offering less in the way of forage than more open and developed land. From the standpoint of deer, Potter County is something of a ghetto in which a lot of individuals are constantly competing for an inadequate food supply. One result is that the Potter deer, though numerous, are fairly scrawny. According to the records of the State Game Commission, Potter is, in terms of deer quality (based on body weight, antler size and reproductive rate) a Class III county. There are 25 Class I and II (excellent to good) Pennsylvania counties in which game managers feel the deer are "better" than in Potter.

To such statistical information the deer-hunting fraternity reacts in a kind of facts-be-damned manner. Potter continues to be what it has been for half a century—*the* deer-hunting mecca. All over the state— in fact, all over the country, because there are those who make the pilgrimage from as far away as Quebec and California—there are sports who regard getting up to Potter for Opening Day as one of life's annual imperatives.

The Hunt as Psycho-Historic Drama

Somewhat like the Masters golf tournament, the Potter County deer hunt has a mighty, all but mythic, reputation. Both reputations are based on a complex of traditions and illusions so powerful that they overwhelm dull, disparaging facts, creating an independent, more or less metaphysical, reality.

On the Sunday afternoon before Opening Monday, on a ridgetop that is the continental divide between the Atlantic and Gulf of Mexico drainage systems, a fellow by the name of Hank Mason is standing in two feet of snow, talking about the intangibles of deer hunting, which is what he will be doing in the morning, and of Potter County, which is where he is. In real life Mason is a plumber from York County, which, located between Harrisburg, Philadelphia and Baltimore, is part of the Eastern megalopolis. There is some remaining farmland, woods and preserved open space around York, but it is well mixed with Interstates, shopping centers and residential, commercial and industrial centers. Despite— and in ecological truth because of—such development, York has quite a few deer, including some very good ones, being one of the state's Class I white-tail counties. Mason might better have stayed home, but he is up in Potter, up to his knees in snow, explaining his preferences and position.

"What it boils down to," he says, "is that I'd just as soon shoot a 120-pound buck, or maybe none at all, up here as a 150-pounder down home." Mason is a burly, crew-cut, exceedingly active hunter, even though he lost a leg during his military service. Among the crowd he regularly hunts with there is a standing joke that they will not have to worry about firewood as long as they have Pegleg Mason. "It's wild enough up here to make you think about how it was when we had to hunt and take care of ourselves—when we were more independent. I suppose that is part of the idea behind those God's Country signs." (For some time tourism promoters have been distributing T-shirts, bumper stickers and other literature asserting that POTTER COUNTY IS GOD'S COUNTRY.)

With its thousands of acres of rough, relatively roadless and unsettled land overlaid by the Black Forest, Potter County looks like a place out of a peculiarly American storybook—Leatherstocking country, the Big Two-Hearted River wilds, Yoknapatawpha North. It is similar enough to the remembered romances to support the illusion of being the Big Woods where, game biologists aside, proper American deer should live and proper American men should pursue and slay them. (At least for Opening Day, God's Country lies far from ERAland. Meeting a female deerstalker is a rare occurrence, a matter for snickers and speculation, like encountering a crowing hen.)

Boiled down another way — and forgetting such practical benefits as exercise, gamy venison steaks or a trophy for the rec room wall — getting out Opening Day offers an opportunity to play a role in one of the most enduringly popular American folk pageants, the Deer Hunt. The part can be essayed anyplace where there might conceivably be deer, but it is more artistically satisfying against a Big Woodsy backdrop. Potter County may not have the highest-quality deer, but it is a Class I set for the Deer Hunt.

Dressing Up

Like Halloween, Mardi Gras and Masonic conventions, the Deer Hunt has become something of a costume affair. As the leaves begin to turn, sports all over the Commonwealth begin to think about what they will wear and carry up to Potter. Red-checked hunting suits are taken out of mothballs and aired. Boots are soaked with neat's-foot and buffed, sheath knives honed. Shells are loaded and the old Savages, Winchesters, Remingtons and Marlins are oiled, sighted and lovingly placed in rear-window racks. Serious performers usually add a few discretionary props, at the very least a flask of booze and a pack of chewing tobacco — not a can of sissified Skoal, but a black, ugly plug of Days Work or a pouch of Red Man, which is sometimes ceremoniously used but more often is kept for hip-pocket display.

Generally deer hunters use far less nylon, plastic and goose fuzz than do, say, backpackers and bird watchers, contemporary outdoor recreationists who, even in the environs of Houston, are heavily into Nanook of the North and Sir Edmund Hillary roles. From the *de rigueur* longjohns, worn even in the balmiest of bluebird weather, up through solid layers of socks, boots, baggy pants, heavy shirts, vests and coats, deer hunters favor garments of moderately archaic style and material, being big leather, canvas and wool men. Right down to the Red Man pouch, the mode of dress is essentially a modern and fairly expensive replica of what is popularly supposed to have been worn by rural manual laborers in the second half of the 19th century, when scratchy underwear, clodhoppers and iron-weight britches were about all that was available in many districts. It was probably no worse an outfit for killing deer than it was for slaughtering hogs, two activities then similar. Getting a deer was not a sport but a protein-gathering chore and was accomplished by turning loose the dogs, setting out baits or traps, doing whatever was necessary to come by meat with as little effort and fuss as possible.

Historical as they may be, 1875 Farm Hand suits make a sporting costume that is almost as cumbersome as those medieval knights wore for jousting. That only the most prime sports can bear up for very far or very long under a 60-pound Deer Hunt costume may be the principal reason so many of them seem to be hunting closer and closer to open roads, a tendency observed and deplored by many game biologists and managers. Probably the most sensible hunting suit of all was devised centuries ago by the Iroquois and other tribesmen who once chased deer through the Potter hills wearing moccasins, flimsy leather briefs and *cagoules*. The modern equivalent would be Nike shoes, Frank Shorter running trunks and a KISS T-shirt. Such garb would be unthinkable for most sports, however, because in it they would not be recognized as members of the Deer Hunt cast and might be mistaken for a Sierra Club freak or Erich Segal.

Occasionally some costumes that border on the surrealistic are assembled out of ignorance or fantastical notions about the theme of the Deer Hunt. Recently Dennis Goodenough, a portly, pipe-smoking entrepreneur who is the proprietor of Potter County's largest sporting-goods store and thus sells and observes a lot of hunting getups, encountered a pair dressed so outlandishly as to give even him pause. "They pulled up, a couple of fellows with long hair and scraggly beards. They looked like radicals you see on TV, and I thought at first maybe it was a stickup, but it turned out they were just crazy, not dangerous. They were wearing safari outfits, and they each had about two yards apiece of cartridge belts wound around them. One had a cow horn, the kind you blow through like Robin Hood, hung on a shoulder. He said a fellow in a surplus store in Philadelphia had sold it to him for calling deer. The same fellow, for $120, had sold them a beat-up old Jap rifle that I swear was not worth $1.20. The stock was cracked, it was covered with rust and the barrel was about as straight as a snake swimming. They'd also got hold of an old four-power scope—the lens looked like it was made out of milk glass. They wanted to buy a clamp so they could put it on that worthless gun. I told them I didn't have anything like that and they should get rid of that Jap rifle before they killed themselves or somebody else, but they went on their way. They said they'd get a piece of wire and tie the scope on with that."

Tod's Kin-Kan Kamp

Along with growing potatoes and cutting timber, tourism is a principal business up in Potter, but it is still something of a cottage industry. There are, for example, only about 240 motel and hotel rooms in the county, less than are found around many intersections on Interstates (none of which crosses Potter). Obviously only a few of the sports who pour in over the narrow approach roads for Opening Day can lodge in public rooms. Through the years, various arrangements have been made for putting up the remaining thousands. The accommodations are mostly primitive, but this has come to be regarded not as a hardship but as one of the attractions of going up to Potter—something that enhances the old-timey, roughing-it, manly theme of the hunt.

A good many hunters wheel into Potter in the modern equivalents of Conestoga wagons: motor homes, trailers or pickup campers. The better class unhitch, so to speak, in approved parking lots, but many others are less choosy, whoaing up their rigs in any likely-looking opening: the entrance to a driveway, farmyards or even in the middle of public roads. "I imagine that they are not such dumb sons of bitches back home as they seem to be up here," a local agriculturist said after having shooed away a pair of 4WD vans whose operators had churned through the mud and snow to set up light mobile housekeeping in the middle of his potato patch. "Up here they see a place without a barbecue pit or flower bushes on it and they think it still belongs to the Indians. Probably they don't even walk across their neighbors' lawns back in Philadelphia or wherever." (Up in Potter, Philadelphia tends to be used generically, as a term referring to the place exclusively inhabited by people who couldn't possibly find their fannies with both hands.)

Other thousands of sports are taken in by local farm ladies who, for $20 or so a day apiece, squeeze a dozen or so hunters onto cots in attics, spare bedrooms and outbuildings and give them three very square meals, allowing them to live much like—and apparently enjoy for a few days the domestic arrangements of—old-time threshing crews. Because for many rural families these visiting sports are, pound for pound, a more valuable and easier cash crop than potatoes, occasional trespassers in potato fields are treated more leniently than they might be elsewhere. This is also the reason so little of the private land in Potter is posted.

More than they do anyplace else, hunters who come up to Potter stay in establishments that might be called cabins or used house trailers or, even more likely, shacks, but which in the Big Woods are invariably referred to as camps, as in Big Buck Camp, Itchy Finger Camp, Good Luck Camp, Shoot 'Em and Miss Camp, Ramrod Hunting Club Camp. There are some 4,500 of these in Potter. Not for economic reasons but out of respect for deer-hunting traditions, most of them are designed and carefully maintained in 19th-century Rural Shanty style, whether made of plywood, tin or unpainted slab lumber. Many of the properties are communally owned by groups of congenial gunners, and membership shares and privileges in these recreational collectives are passed along from father to son.

In several senses of the word, a splendid example of a Potter County deer camp is one situated in a hemlock-filled ravine through which flows a streamlet called Meeker Run. It is the sporting headquarters for the male members of a family whose roots are in Adams County, in which Gettysburg is located, 150 miles to the south. The Meeker Hollow camp was organized in 1937 by the four Raffensperger brothers and their father. Two of the brothers—George, who is 80 and a retired banker, and John, 68, a retired fruit grower—continue to hunt from the camp but now share it with sons, sons-in-law and grandsons. Some of them have scattered across the country, become farmers, college professors and businessmen, but they make a considerable effort to get back to Potter for Opening Day.

At first, the Raffenspergers rented the land along the run and each year would put up a large canvas tent. After World War II they bought 12 acres around the tent site and a heavy, 32-foot-long steel-bodied truck trailer. This they hauled up to Meeker Hollow as the shell of their camp. The trailer was set up on the property with some difficulty. Beyond the obvious problems of fixing it on the side of the ravine, the brothers were determined to save a fine maple tree that now shades the camp. "When we tented," says George Raffensperger, "we used that tree to tie to. It wasn't much more than a sapling then. We wanted to save it as a sentimental reminder of the old days."

In memory of their father, Tod, and as a jest about the material the camp is made of, the brothers call it Tod's Kin-Kan. Esthetically the Kan is not impressive, but functionally it evinces years of loving and cunning craft, an accretion of improvements and conveniences added by various relatives. Eight plank bunks fill the north end, the lumber worn smooth through the years by wool blankets. At the other end there is a chunk-wood stove for heating and an electric stove for cooking; also a plank table and benches. In between is a traditional peg, shelf or niche for every pot, pan, gun case, hunting coat and hat. Everything is neat as a pin, not as neat as a wife would keep a place but rough-neat, as good logging or cow or mining camps were once kept by men. There is a pleasant nostalgic smell of firewood, flapjacks, bacon, wool and boot and gun oil about Tod's Kin-Kan.

The good book of the camp is a log in which events of each hunting season are meticulously recorded: the weather, who was there, interesting sightings of foxes, mink, owls, hawks, and, of course, who took what kind of deer where and when. All told, the Kin have killed about 100 bucks working out of the Meeker Hollow camp. The last day of the season is always set aside for butchering. The venison is divided equally among the members of the party, with the man who actually brought down the animal also receiving the head and hide. It is a matter of pride among the Kin that during 40 years of hunting none of them has violated any Pennsylvania game law or, for that matter, the unwritten code of good sports hunting—for example, that wounded deer must be tracked down and dispatched.

"This week is a reunion for the men in our family and truthfully, a very fine kind of escape," says Edgar Raffensperger, who now comes to the Kin-Kan from Ithaca, New York, where he is a professor of entomology at Cornell. "Whatever we have been doing, we put it aside when we come up here and become totally involved in camp work and hunting, in older ways of doing and living and thinking."

Whether it is more fun for bankers and professors to play golf or to play at being 19th-century woodsmen is, of course, entirely a matter of individual preference. However, the party at Tod's Kin-Kan Kamp is an impressive and admirable recreational group. If everybody were like the Raffenspergers, the phrase "slob hunter"—for that matter, just plain "slob"—would never have been invented.

Rendezvous

In 1822 a sometime soldier and Congressman and all-the-time hustler from St. Louis by the name of General William Henry Ashley cajoled a group of unemployed farmhands into heading off to the wilds of the Rocky Mountains to trap beaver. Thereafter he devised a system for getting his hands on the furs. He called it the "rendezvous." Late each spring Ashley would send a packtrain out from the Missouri settlements loaded with trappers' supplies and a lot of raw rum. This caravan would reach a predetermined riverbank meadow at the base of the mountains, down from which would come the trappers, literally loaded with beaver pelts and figuratively loaded for bear. Ashley's agents would purchase the pelts—at rock-bottom prices, because they were the only buyers within about 1,500 miles. Then, to take the sting out of the transaction, they would break open the rum kegs, the marked decks of cards, and urge the mountain men to let it all hang out. By report, not much encouragement was needed. For the next few weeks they drank steadily, gambled recklessly, pushed, shoved, wrestled, bit ears, gouged eyes, bragged about the hair on their chests, played dreadful practical jokes, like throwing rabid wolves into each other's tents, and generally whooped and hollered. When it was all over, the trappers were left drained, dead broke and in hock for part of their next year's catch. Ashley's agents headed back to the bright lights with all the furs, most of the money and lots of amusing stories about the excesses and stupidities of the mountain men.

As an effective device for economic exploitation, the rendezvous lasted only about a decade, or until the beaver were all gone. However, as an entertainment (which was widely publicized, first by word of mouth and later in pennydreadful novels and all manner of subsequent media forms), the rendezvous permeated the imagination of American males and, it seems, had a lasting effect on their behavior. Though the original facts have become distorted and by now are all but forgotten, the legacy of the rendezvous is that when a group of American men find themselves far from hearth and home, for various vague, chauvinistic and machismo reasons they are more or less required to have themselves a bash. The spirit of William Henry Ashley's rendezvous lives on—at fraternal and business conventions, class reunions, in locker rooms and military and civilian R-and-R centers, and especially on hook-and-bullet outings.

The good steady hunter ethic prevails in many places like Tod's Kin-Kan Kamp, but a number of those taking part in the Deer Hunt feel obliged to do a few rendezvous turns when they arrive in God's Country. A good place to observe this action is a public watering place, of which there are many more in Potter County than there are motels, bowling alleys and movie theaters. One of these can be misnamed (for reasons of propriety and back-home peace of mind) the Roaring Buck, which is more or less what a lot of spots are called in Potter. On the Sunday night before Opening Day, 50 or 60 flatland sports are pushing and shoving in the Roaring Buck, drinking and sweating. Everyone is already dressed up in longjohns and wool suits and talking very big. Great stretchers are told about the size of bucks that have been spotted and will be shot come the morning; about how much snow has been wallowed through; about long-range shots of the past; about what they said to a dumb so-and-so who tried to take a deer they had downed. In the great American rhetorical I'm-half-alligator-half-wolf- and-was-born-in-a-grizzly-den tradition that goes back at least as far as the original mountain men, there is a lot of stylized bragging, but as a reflection of the times, it is not so much about personal prowess as about the powers of mechanical possessions—scopes, 4WD rigs, CB transmitters and snowmobiles.

One of the merriest groups in the Roaring Buck is made up of half a dozen men who come from the eastern part of the state, where they are executives and managers of a large sheet-metal firm. They have been hunting together up in Potter for almost a quarter of a century and, for obscure reasons of their own, call each other, at least during this outing, Cousin, as in Cousin Ed, Cousin Willy, Cousin Flunky. A prominent cousin and sales manager is Sam, who among other distinctions has never shot a buck (in fact has only shot at one on three occasions) though he has been trying to do so for 22 years. Cousin Sam's bad luck has become a folk myth for the group, and he takes a lot of needling about it. ("What he does not tell," howls Cousin Ed, "is that he very seldom gets a chance to see a deer because he is too busy taking our money at poker.") However, he remains the most exuberant of the cousins, belting back the vodka and Squirt at a great rate. Also, he is carrying on in a very public fashion with a blonde waitress by the name of Joan. In regard to Joan, Cousin Sam has two things in mind, or so he loudly keeps saying. He would first like her to sit on his lap—which she occasionally does, perching for a few seconds on his knee like a bird and leaving as easily and quickly. Secondly, he would like her to inspect his brand-new $26,000 motor home, which is parked outside the Roaring Buck. She turns down this invitation, but in a manner that doesn't spoil the fun.

In the days of the first rendezvous there were always a few girls in the mountain meadow, mostly Snake and Crow beauties, but arrangements about them had to be made with hard-bargaining male relatives, who brought the women along more or less as trade goods. That is how it always has been at rendezvous—very few, very inaccessible women and a lot of men who have removed themselves from regular centers of femininity but who talk stud incessantly. At the Roaring Buck it would seem, just from eavesdropping, that virtually every man jack has just come from or is just leaving for an assignation.

Joan is, in fact, a young woman who has left a Pennsylvania college for the fall semester to repair her finances by working at the Roaring Buck. However, for the moment she is in the historical position of a Snake woman, and her views about rendezvous may be quite similar. "When I see hunter's orange I think of green bills," she says, taking a break behind the kitchen door while waiting for a hard-pressed bartender to catch up on back orders of vodka and Squirt. "Last night that ———," she nods toward Cousin Sam, "put a 20 in my pocket. I guess he was trying to buy some respect. Did you see him?"

"What happened?"

"He got bragging about chewing tobacco. He stuffed half a pack of Red Man into his big mouth and threw up his whole dinner.

"Old Cousin Sam is a perfect example of a guy whose wife lets him out about three times a year, and he tries to act like he does it all the time. He's a nerd but absolutely harmless. Some of them are mean. You wouldn't believe the black and blue marks I've got!"

The Shoot

By and by Cousin Sam gets into the motor home, without Joan, and weaves back to the farmhouse where his party always bunks. He cannot sleep—"That cot was spinning around just a little too much"—but drinks a lot of coffee, lays out some solitaire and is inspired to play a good practical joke on Cousin Flunky. He sneaks Flunky's boots from under the bed, ties the rawhide laces together in double granny knots, wets them and puts the boots in front of the fire to dry. "We all split a gut laughing at old Flunky when he wakes up hung over and tries to pick out those knots."

At 6:43 a.m., which is precisely when the laws of Pennsylvania permit hunting to begin in Potter, Cousin Sam is walking down a snow-covered woods road behind the farmhouse, carrying his Ruger across his arm. He follows the road a few hundred yards and then stops in the shelter of an oak in a spot that commands a view of an open ridge above. "I didn't see any point in wearing myself out in that snow," Cousin Sam said later. "I figured I'd save my strength and let the other boys get the deer moving."

About 8:30 six deer appear on the ridge 100 yards or so away. "I'm pretty sure one of them showed some antler [a buck must have three inches of spike or a fork antler to be legal quarry], but I wasn't dead sure, and then I got thinking if it was a spike and I did hit it I'd have to wallow through all that snow to get it and drag it back to the farm. I said to hell with it, let it go, and kept my record intact."

By five p.m. on Opening Day, Cousin Sam has had a large lunch at the farmhouse, a refreshing nap, won $64 in the afternoon stud game and is back at the Roaring Buck and the vodka and Squirt. "Some of them may get more deer," he says jovially, "but nobody has more fun hunting up in Potter than I do. Who needs a stinking old buck when you got a sweet little doe around,'' and he makes a slow, unsuccessful lunge at Joan.

Things go almost as predictably but much differently at Tod's Kin-Kan. "Once in a while somebody drinks a bottle of beer," George Raffensperger says, "but we're not much for playing cards or staying up late. I guess we were all in bed by nine. We save our fun for daylight."

By daylight the Kin are up, have done their chores around the Kan, and are well back into the woods. "I worked my way up on the ridge above that section the timber company has clear-cut," says John Raffensperger that evening. At 68 he moves through the rough terrain with ease and assurance. "I'd scouted the day before, and I could tell from the sign that deer were moving there." Around 10 in the morning three deer show up about 120 yards below Raffensperger. Motionless and downwind, he waits until the animals settle down and he can determine their sex. Then he drops a seven-point buck with a single chest shot. As required by law, he fixes a big-game tag on the buck, efficiently field-dresses it and drags the carcass down the ridge, across Meeker Run and up the far side of the ravine to a secondary road.

"I guess the years are catching up," he confesses. "I was puffing by the time I got back to the road. A young fellow from one of the other camps in the Hollow came by in his truck, and we loaded in the buck and drove it back here. I tried to give him $2 for his trouble, but he wouldn't take it. You meet some awful nice fellows up here."

Gary Horst, a 22-year-old steel welder from the Lancaster area, had one of the more active Opening Day hunts. Horst is an all-round Potter buff, coming up regularly to God's Country to ski, snowmobile and fish, usually staying at the Potato City Motor Inn on the edge of the Black Forest, the largest resort in the county. He is also a deer hunter, but, as he admits, a fairly green one who in four previous seasons had been skunked. "Something always happened. Either I wouldn't get a shot, or I'd have one and choke—miss it. Once it was the old excuse, a shell jammed."

After a bit of rendezvousing, Horst was up at 4:30 on Opening Day and got a Thermos of coffee and pack of sandwiches from the hotel kitchen. He drove his Jeep up Dry Hollow Road, parked and worked his way back into a white-pine plantation. At eight, three does and a spike buck appeared about 50 yards in front of him.

"I started shooting as fast as I could," Horst says. "I knew I hit him someplace, but he kept flying toward the pines. Then there was a click. I'd fired the whole clip, five shots. I reached in my pocket for more shells, and my hand came out all covered with some thick, sticky yellow stuff. I was so excited I thought at first I'd hurt myself, but it was egg salad from my sandwich. There I was with my mittens off, my hands freezing, trying to scrape egg salad off the shells. I dropped most of them in the snow but I got two cleaned. By that time the deer was gone but I was damned if I was going to let him get away.

"There was a blood trail, and I started following. I followed over three ridges for about an hour. I caught up just as he was going in some more pines and shot again, but he still didn't go down. I had only one shell left, so I went above the pines, which looked like the way he was heading. He came out 30 yards away and I shot him in the neck. He staggered into a hollow and died there. I had to figure out how to gut him by trial and error because I'd never done it before. Then I dragged him about a mile and a half to the road and from there walked back to get my Jeep. After I got him loaded in I remembered I had dropped my Thermos and binoculars in the snow where I first shot, so I hiked back for them. It took me about five hours to get that deer, and when I got back to Potato City I just flopped down. I was dead to the world for three hours."

Horst hung the deer behind the hotel. Dressed, it weighed about 90 pounds and the carcass was considerably mangled. As it turns out, Horst hit the spike buck on four out of seven shots—in the rump, belly, neck and shoulder, a shot that had all but severed the right foreleg. "When you think of it," Horst says, "that was one tough little deer."

The Law

Dick Curfman is the Pennsylvania game protector in the northwestern section of Potter County. As such he is more or less the referee for Opening Day in those parts. It is a job that makes any other sort of sports officiating—even an extreme case like refereeing a North Carolina–NC State basketball game—look like very small potatoes, as they might say in God's Country. To begin with, hunters are fully as capable of committing violations—and just as tricky about it—as any performer in Reynolds Coliseum. Last year 10,942 sports were nailed for game violations in Pennsylvania. On Opening Day, Curfman tries to keep his eye on some 10,000 participants, who are maneuvering about in 400 to 500 square miles.

Because all living things, except those co-opted by the feds, legally belong to the state, Curfman is responsible for protecting every wild animal, from bats to bears, in his district. However, in practice he admits that what with poaching of deer, motorists running into deer on the roads, deer running into vehicles, deer ripping off potato fields and orchards, a good part of his time during the course of the year is taken up with this one species and its encounters with people. Opening Day is, of course, all deer, and it actually begins for Curfman late Sunday night when he gets a call from an irate landowner about a suspicious truck in his pasture. Curfman eventually apprehends two sports trying to get a big jump on the others by jacklighting a buck. Then, after a few hours' sleep, he sets out on the day-long patrol, accompanied by his 12-year-old son, Rich, a student-trainee by the name of Mark Crowder and a curious civilian observer.

In the course of Opening Day, Curfman averages about one violator every half hour. A good many of the infractions are routine. Half a dozen citations are passed to hunters who are found with loaded guns in their vehicles—a regulation designed more for the protection of people than game. An equal number of sports fail to fill out and affix tags to their kills, a bit of paper work that is required in an attempt to thwart the ambition of a lot of hunters to shoot more than the one-deer limit. Other sports are hunting from the road, which is also illegal, and a pair of them are caught doing a little creative carving—trying to manufacture a spike antler for a deer they have shot that did not have one of its own of legal length.

The last pinch of the day occurs well after dark, an hour or so after Opening Day has legally ended (at precisely 4:49). On an isolated forest road Curfman comes upon two corpulent hunters who by the lights of their truck are trying to drag a deer out of a ravine.

"Jeepers Jenny, that's a nice buck you fellows have, but it's getting awful late to be fooling around in these dad-gummed woods," says

Curfman, who tends to come on very softly, like a rural Columbo, and also has a great talent for creative, non-profane expletives.

"We've been dragging through that —— snow for an hour," the spokesman hunter replies. He is a little testy, a state which can perhaps be excused because he is a very plump man, sweating profusely in full Deer Hunt costume.

Curfman nods sympathetically and bends down to look at the buck. "Holy Tomatoes!" he exclaims, apparently more in sorrow than anger. "You fellows forgot to tag it."

The lead sport turns righteously indignant. "Look, it was dark. We lost our pencil someplace. We were going to do it when we got back to the truck. I don't see any reason for you harassing us. My name is Ralph Ruckus and I'm a police officer myself in Frozen Pizza, New Jersey [or somebody from a place like that]. Where I come from there is such a thing as courtesy between fellow officers."

"Oaky-Doaker," Curfman says amiably. "Now let's see—this is your buck, Ralph?"

"No, he shot it." Ruckus nods to his companion, who seems to be suffering from considerable nervous tension.

"Sal Banana," the second man introduces himself.

"Jeepers, that must have been quite a shot, Sal—lot of thick hemlock up this hollow. How far in were you?"

"Well, I don't know exactly. A long way, a quarter of a mile."

"Farther," breaks in Ruckus. "At least a half mile."

"You must have got him about four o'clock," presses Curfman.

"Probably earlier than that," says Ruckus emphatically.

"But Jeepers Jenny, you said it was dark when you lost your pencil. It's been a nice clear day. The sun didn't set until four-thirty."

"Well it was getting dark back in those —— woods."

"Oaky-Doaker. I guess I'm going to have to write you up for no tag," says Curfman, and both men look relieved. However, before he brings out his citation pad the game protector bends over the deer again and pats the carcass, which is steaming noticeably. "Nice buck, but Holy Tomatoes, he sure is warm for having been dead so long."

"Look," Ruckus blows up. "I'm tired of these insinuations. What the hell are you accusing us of?"

"Oaky-Doaker," says Curfman, and there is a click in his voice like a bolt being shoved home. "Since you brought the subject up, I'm just going to have to tell you. But first, Mark and Rich," Curfman says, turning to his student-trainee and his son, "why don't you follow those tracks back into the woods a little way. You might find the pencil they lost, and you just might find something else. Now to get back to your question. I feel this deer was shot right here on the edge of the road. Also, judging from the deer's eyes and the body temperature, I'm positive this deer has been dead at most a half an hour, which is an hour after it is legal to shoot. Now if you two fellows have anything you wish to tell me before Rich and Mark get back, I'm willing to listen to you."

Only about 50 yards back in the hemlocks Mark and Rich find the kill spot and a pile of warm deer entrails. However, by the time they return to the road, Ruckus and Banana have come clean and confirmed Curfman's suspicions. It costs them $50.

Curfman has been a game protector for 19 years, and last year he was designated as Pennsylvania's outstanding game officer. In the course of his service he has seen just about every misdemeanor and felony man can commit against deer and heard some very ingenious excuses and explanations for doing so. Generally he feels things are getting worse. "The quality of the hunter is definitely deteriorating," he says. "There are very few persons today who are willing and able to put on an all-day drive, which is a good way to hunt and good exercise. Each year they seem to stay closer and closer to the roads. A number of them are all too willing to commit a violation. They do things up here they'd never do at home, and, Jeepers Jenny, we only catch a few of them. As far as I'm concerned it's part of the overall disrespect for authority and all forms of law enforcement—that, and that people are forgetting good hunting traditions."

Odo

There is another absolutely essential, if presumably involuntary, participant in the Deer Hunt, *Odocoileus virginianus,* the white-tail. Generally we think about the animal in a composite way, in terms of population, age structures and curves, average annual consumption of food, reproduction, growth and mortality rates, the dynamics of the herd. However, it is well to remember that a "herd of deer" exists only in the same sense that the "upper middle class" exists. It is an abstraction, a convenient way of dealing with a large group of somewhat similar creatures, each of which, by reason of genetic and existential distinctions, is as unique as, say, John Raffensperger, Cousin Sam and Dick Curfman are unique.

Consider, so far as it is possible, one of these unique individuals of another species who is in Potter County on Opening Day. Call him Odo for short. Most obviously he is set apart from his kind by reason of his size and condition, being a buck of some 160 pounds, carrying a rack with a 20-inch spread and 12 points. He may be a 5-year-old, and if so has reached an extraordinary age, not in terms of the natural life-span of the species, which is up to 20 years, but in terms of bucks in Potter County, where about 75 percent of all antlered deer are killed each year during the hunting season. Odo may well be one of the very last of his class of several thousand male fawns born in the spring of 1972.

The presumption is that he was born close to where he is on this Opening Day, in the valley of Bailey Run in southern Potter County. Deer are not great travelers, and Odo probably spent most of his five years in the same square mile, following like an unadventurous commuter the same trails he found and made as a yearling, traveling them for purposes of foraging half a bushel or so a day of hemlock, tamarack, cherry, maple, oak, blackberry, grape, blueberry, wild-rose shoots, twigs, acorn mast and apples from several abandoned orchards in his territory, as well as other aquatic and upland herbage. Again because of his condition, he has apparently fed well and been fortunate in avoiding serious infestations of tapeworms, liver flukes, lung worms, botflies, ticks and mites. He may have survived and prospered simply by chance, or perhaps because of some mysterious superiority of the eye, nose, muscles, glands or brain.

It is likely that he has bred several, perhaps half a dozen, does each of the past three autumns. If so, he has ferociously fought for their favors against other bucks, probably successfully, given his strength and size. Early on the morning of Opening Day he is with a doe, perhaps one he had mated with the previous month. However, by Thanksgiving weekend the great hormonal tides that aroused him during rut have subsided. His once tawny coat has turned darker and the narrow ridge of bone below the coronet of his antlers has begun to be absorbed, preparatory to the dropping of the rack. His reasons for being with the doe are other than reproduction. They may be together for companionship or security or communal pleasure, but this is sheer speculation, because when it comes to intraspecific behavior, we are knowledgeable only about the grossest relationships.

Odo and the doe have been nibbling, seeking shelter from the snow in a grove of hemlocks in which they normally spend the daylight hours. However, this is an unsettling morning, with the sound of gunfire and disturbing scents in the woods. To escape the alarms the doe leaves the thicket and starts up a transridge trail. Odo follows. When he comes into the open he halts, testing the air, apparently trying to locate and identify a strong and suspicious foreign scent. In that second or so there is a roar from a rifle, and almost simultaneously a bullet from a .30-06 Ruger smashes into his chest.

What does he feel? Again this cannot be precisely documented, but because Odo is a mammal with a highly developed nervous system, he probably feels much as the man who pulled the trigger would if by some chance the buck were to gore him through the chest with an antler. From testimony of our own kind who have suffered and survived somewhat comparable experiences, there is a sensation of ripping and tearing as the bullet shatters bone and muscle, then terrible burning pain in the destroyed lungs. In agony and shock Odo makes a final instinctive lunge, but the marvelous legs can no longer carry him through the woods at 40 mph. The wound is too severe. He drops on the spot and dies in the snow above Bailey Run.

From that moment on, by custom and law, Odo, or at least his carcass, belongs to Robert Stahlman, a 31-year-old steelworker from Warren, Ohio, who fired the Ruger. Stahlman is immediately aware of, and ecstatic about, the size of this buck. After tagging and gutting the body he takes it to Dennis Goodenough's sporting-goods store in Coudersport, which is the official headquarters of the annual Potter County Big Buck contest. Eventually Stahlman and Odo win it. The rack scores 32 trophy points (on the basis of antler size and spread). When the season is concluded, Odo is certified as being the best buck shot in Potter County. While the measurements are being made, the carcass hangs outside

Goodenough's store, and Stahlman hangs around inside accepting congratulations, smiling and bursting with pride.

"I'd have been happy with anything," he says, "even a spike. When I saw that monster I couldn't believe it. When he went down it was like the big moment in my life, like winning some big game. I never thought anything like that would happen to me. I just thanked God for that buck and that I was there and got that shot."

"I guess you felt like God had smiled on you in God's Country."

"Right—that's just how it was."

Baffinland

Baffin is a 1,000-mile-long island, the fifth-largest in the world, and is shaped like a crude, badly used battleax. The notched, dented blade, two-thirds of which lies above the Arctic Circle, partially caps the northern reaches of Hudson Bay. The shattered, stubby haft, a promontory called the Cumberland Peninsula, points toward Greenland, 250 miles eastward across the polar seas.

By any standards, Baffin is a formidable place. The average *maximum* winter temperatures are well below zero. If summer comes — and often it barely gets to parts of Baffin — the midnight sun may warm things up to 50° or so, but there is no day in the year when there is not the possibility of frost and snowfall. Winds are incessant and often of gale force.

The island's terrain is rough everywhere, but it becomes truly mountainous on the eastern side. The elevation of the black, rocky peaks there, five to seven thousand feet, is not spectacular, but their conformation is. Geological upheaval, ice, water and wind have cut the rock into jagged spines, columns and abutments, the sheer faces of which rise several thousand feet straight into the cold air.

According to old Baffin hands, God created the rest of the world in five days, on the sixth He made Baffin and on the seventh He amused Himself by chunking rocks and ice at it.

Much of the land that is not rocky is icy, lying under banks of perpetual snow and glaciers. The massive ice fields are the remnants of the Laurentide ice sheet that once covered much of Canada and portions of the U.S. to thicknesses of up to 13,000 feet. The Laurentide ice eventually retreated to Baffin Island, where it probably began. It waits now in the black mountains, the presumption being that sometime it will once again move south.

Though in area it is larger than New England, New Brunswick and Nova Scotia combined, the total population of Baffin could fit into the bleachers at Fenway Park. This is not surprising, given the grim nature of the island, but what is surprising is that people have been trying to live there since long before they began to nose around Boston Common. In the course of their epic migration across the Arctic, Eskimos reached Baffin some 4,000 years ago and have remained ever since. Some 5,500 Eskimos (or Inuit, as they call themselves) live on the island, most of them in half a dozen recently established coastal villages that are convenient for government administrators, social workers and storekeepers up from southern Canada.

About 1,000 years ago Vikings came to Baffin, but while Norsemen continued to stop by periodically for centuries, there is no evidence that they enjoyed or profited from their visits. In 1585 a British mariner, John Davis (Davis Strait, between Baffin and Greenland, is named for him) made a landfall on the Cumberland Peninsula. Thirty years later William Baffin explored the island, and since then there have always been whites in the area—first whalers, then traders, and currently mostly bureaucrats of one form or another.

About 20 years ago, members of the international rock and mountain climbing fraternity began to investigate Baffin, particularly the strange peaks and extensive ice fields of the Cumberland Peninsula. Now this arctic wilderness has become for very serious rock climbers and backpackers what an ultrafashionable disco is to the international jet set; in other words, the appetite for exotic and adventurous recreation that has developed in the affluent temperate zones has made Baffin a place to be.

Like rock dancers, rock climbers come from all over the more or less civilized world. Among the several hundred who made it into the interior of Baffin in the summer of 1978, for example, were Scots, English, French, Luxembourgers, Swiss, Japanese, a Korean and a scattering of North Americans from such places as Fairbanks, Alaska; Vancouver, B.C.; New York City and Iron Springs, Pennsylvania.

Six years ago, in part to both aid and supervise outside visitors, Canada set aside 8,287 square miles of the Cumberland Peninsula as a national park called Auyuittuq, an Inuit word meaning The Land That Never Melts. Among other impressive phenomena, the park includes the Penny ice cap, one of the world's largest sheets of permanent terrestrial ice.

Auyuittuq is the northernmost park on the continent. A pass climbs up the valley of the Weasel River from the head of a fjord off Cumberland Sound. It reaches a series of glacial lakes at the summit and then descends the Owl River to an arm of Davis Strait. The route along the Weasel (but not along the Owl) is sporadically marked by Inuit cairns. Otherwise there are no signs or guideposts. There are seven tiny emergency shelters into which three or four people can squeeze to escape the weather. This is the extent of the facilities in Auyuittuq.

On the fjord below the Weasel River, 18 miles from Auyuittuq, is an Inuit village of 950 people called Pangnirtung. In it are the park headquarters, where maps and advice can be obtained. The advice is largely cautionary. Visitors are given to understand that if they get into trouble in Auyuittuq they will have the sympathy of the park field staff (four wardens and a superintendent) but not much else. No rescue teams are available to respond to emergencies. The point is emphasized that anyone who goes into Auyuittuq must be self-contained and self-reliant.

There are some gaudy ways of entering Auyuittuq, such as parachuting onto the Penny ice cap, but most people fly into Pangnirtung and there try to make arrangements with Inuit guides to be hauled up the fjord in canoes or on snow sleds. Even getting to Pangnirtung by more or less conventional transport can involve considerable logistic effort, and when you get there you wait. You wait for machines, for spare parts, for somebody's cousin who is said to be seal hunting but who, it is also said, will probably be glad to give you a ride in his boat if and when he returns. Especially you wait for the weather—until it is good enough to fly or paddle or just walk.

The problem of arctic travel was nicely illustrated in Pangnirtung last July, which is when summer is supposed to occur. Gales, snowstorms and frosts were frequent and the weather was generally worse than anyone could remember it for 20 would-be summers. A few impetuous visitors who got to Pang early were able to snowmobile into Auyuittuq over the frozen fjord, but shortly thereafter the softening ice became too rotten for sleds while not dissipating sufficiently to permit canoe travel. These in-between conditions continued almost until August. The *crème de la crème* of the international outdoor set, those who had arrived precisely when careful research and planning indicated they should arrive, hung around Pangnirtung for days and weeks while their supplies, patience and cash melted. Waiting in the Arctic can be as expensive as traveling. In Pang, lettuce was selling for $3 a head, apples for $1 each. Other goods and services were comparatively priced.

Scenically, Pangnirtung has much to recommend it, at least for a few days of waiting. There are those who feel it is the most attractive village site in the Arctic, standing as it does on the beach of a clean, deep, mountain-bordered fjord. But many people, especially romantics who want igloos or, at the least, tents, find the village less than charming. In design and function, Pang is arctic modern, like most of the settlements in far northern North America. The buildings are basically prefab tin and wood boxes, to which additions of packing crates, plastic and canvas are often made. Most of them are surrounded by truly impressive piles of debris—cans, bottles, plastic wrappers, old bedsprings, broken toys. Remnants of outboard motors and snowmobiles are mixed in with exotic organic matter, scraps of fox and hare skins, pieces of old whale and seal, and lots of very defunct fish.

There are sound reasons for this mess. Temperate zone waste-disposal systems are simply out of the question in settlements perched on solid rock or on a few inches of frozen sand over the rock. For thousands of years the obvious solution has been to let things lie and rot where they fall. But because of greater contact with southern civilization, the north now has more junk than it used to but the disposal system remains traditional. A happier way of looking at Pangnirtung is that it is a kitchen midden in the making and no doubt will be very attractive to archaeologists who come this way a thousand years from now.

For tourists, a very important person in Pangnirtung is Ross Peyton, a gaunt Newfoundlander who came to Baffin nearly 30 years ago as a Hudson's Bay Company clerk and stayed on to become an entrepreneur of the Arctic. From his base in Pang, he trades in native crafts and operates a fairly luxurious fishing camp that caters to affluent anglers from the south. Peyton also owns the only public house in Pang, a combination hotel–dormitory–dining room, which is of course known as Peyton Place. Like other more or less permanent inhabitants of Pang—Inuit and southern Canadian alike—Peyton has not done any thrashing around in the interior of Auyuittuq and expresses no interest in doing so. However, he thinks well enough of the park and its users, though they put little money in his pocket.

"The chaps with the big packs," he says, "bring everything with them. They camp out on the beach or up in the rocks, more power to them. They will slip in here when things get grim for a shower or a meal. Generally they don't have much money, but they are very bright and well educated. When they go back they talk a lot about what they have done, and it gets our name around. Older people who don't want to scramble around in the rocks but who have the wherewithal for a more comfortable trip hear about us and may book in the hotel or at the camp. Those pack people do my advertising for me."

More efficiently than any government regulations, environmental imperatives control the number of people who use Auyuittuq. The difficulty of reaching even the head of Pangnirtung fjord winnows out the casual tourist. Only about 1,000 arrive in Pang each summer with some interest in seeing the park. About half of these make it to the park proper but turn back after a one-night stay just inside the Auyuittuq boundary. Only about 250 continue on up toward an emergency shelter 10 miles away at the foot of a gray, cold body of water called Windy Lake. Perhaps 200 of these go on toward the summit of Pangnirtung Pass. Brief guides to Auyuittuq note: "Arctic hiking is quite a bit different and more difficult than walking through other areas of the Canadian wilderness." And: "Needless to say, previous backpacking experience and good physical condition are prerequisites. Novice and intermediate hikers should be accompanied by more expert backpackers."

The walk up the Weasel to Windy Lake is relatively easy compared to what lies beyond, but even in this 10-mile stretch there are "differences and difficulties." The ascent is moderate, the summit of the pass standing only about 1,500 feet above sea level, but the footing is atrocious. The best of it is found along the river on sand and silt bars and in low-lying pockets of mossy, swampy muskeg. The ground squishes up underfoot but at least the walking is fairly level. However, there is not much of this flat, sloppy going because the terrain is choked with moraines, those great haphazard piles of debris left behind by glaciers. Moraines are filled with rock, everything from gravel to boulders the size of suburban bank buildings, and they are slow and fatiguing to cross. The worst are the newest ones, made slippery by residual mud and ice. Alongside and through the moraines run powerful, milky colored streams fed by the melting glaciers above.

The quickest and most tempting way of crossing moraines is to jump from one icy rock to the next, but if you're carrying a full pack this is also a good way of breaking a leg or cracking a skull. A variant technique at difficult points is to change from boots to sneakers and wade up the streams. This is slow and also painful, because the glacial waters of Auyuittuq are a few degrees below freezing, cold enough to burn like fire after a short period of exposure. Two or three minutes would be the average survival time if one were fully immersed in the water.

Aggravating all the other difficulties is the weather. Says the cautionary Auyuittuq literature, "Strong wind combined with rain or wet snow at near the freezing point—conditions which can occur in every month of the year—has caused more exposure casualties than full winter cold." The quality and quantity of the wind is hard to imagine. There is almost never a truly calm day in Pangnirtung Pass, and gales of 40 or 50 mph are as common as zephyrs in the south. These are winds that cut and bully and can kill. According to a clinical note on a topographic map of Auyuittuq, "Hypothermia is the rapid, progressive mental and physical collapse accompanying the chilling of the interior of the human body. It is caused by exposure to cold and aggravated by wetness, wind and exhaustion. Most cases develop in air temperatures between 1°C. and 10°C."

Fortunately, Sam Walmer, the aforementioned Pennsylvania orchardist-porter, was able to leave his Red Delicious plantation to make a trip to Baffin, Nature's own cold storage facility. For a variety of temperamental reasons, Sam and I tend to mount disorganized expeditions. We had fully intended to reach Auyuittuq in mid-July, the optimum time according to all authorities. However, there were distractions. For example, several days were lost in Toronto as we tested the theory that it is easier to handicap horses there than it is in Charles Town, West Virginia. This proved to be untrue. We did not reach Pangnirtung until nearly the end of July—and were rewarded for our tardiness; we caught one of the first canoe flotillas going up the fjord, a happening that better-organized tourists had been awaiting for weeks.

Among our upward-bound companions were four men who identified themselves as public-school teachers of outdoor education. They were abundantly and cunningly equipped and gave us a lot of good tips about how to survive in the outdoors. We hoped that their concern for us was misplaced, but we could understand why it arose. Sam and I do not appear very smart when we go a-venturing. Sam, for example, no matter what the occasion—running white water, spelunking, climbing glaciers—favors a basic costume of overalls, ragged sweat shirt, long johns, a tractor driver's cap and Sears work boots. Our gear tends to be old and battered, having been beat up in a lot of improbable places, including several in the Arctic. Because of sloth, we are both disinclined to carry anything we are not going to use or use up. This suits us, but it is not surprising that *au courant* outdoor experts would worry about us.

When we beached at the head of the fjord the outdoor educators immediately said goodby and added that they might see us again when they were coming down from the summit, at a time when we presumably would still be trudging up. Their plan was to make the 70-mile round trip in four days, and they started off briskly. We dawdled. The day was a nice one, one of the first summer days of the year. Temperatures were in the 40s, there were occasional periods of sun, and the wind was gentle, no more than 20 mph. We found a patch of dry-meadowy muskeg covered with a lot of nice-looking boreal flowers. We set up the tent behind a sheltering boulder and spent the rest of the afternoon admiring nature.

Sometime during the long twilight that passes for night in these parts the weather changed and returned to normal. The temperature dropped to freezing; it began to rain or snow or something in between, and the wind rose and began beating on the tent. There being no incentive to lounge around in this sort of thing, we packed up and started ascending the Weasel River Valley. We reached the Windy Lake shelter about midday and in it found two of the outdoor educators. One had a sore knee and the other was trying to mend a tent that had been ripped by the wind. They said their two companions were scouting ahead to see whether or not they should go any farther.

Windy Lake is descriptively named. All the gales of Auyuittuq appear to fancy this spot as a permanent home. Beyond the shelter we approached the shallow, perpetually roiled lake over a series of long, steep, slippery moraines that in places could be crossed only by crouching low and hanging onto the rocks so as not to be blown backward. In God's good time the moraine gave way to several miles of mud and sand flats over which the unimpeded winds from the lake get in some of their best licks. At the head of the flats, we met the other two outdoor recreationists coming toward us. They had gone a mile or so farther, found no place where a tent could be pitched and said they could not believe how bad the weather was. They had given up the idea of a dash to the summit of the pass and were going back to their two friends, then returning to Pang as soon as weather permitted.

"At least you will have good lecture material when you get back to class," said Sam, with a certain innocent venom.

There was not much fight in the outdoorists. Their leader, a principalish man with a very teachy conversational style, nodded glumly at the environment and said earnestly, "This is a harsh land of many contrasts."

Every expedition, well ordered or not, needs a zingy slogan to lift the spirits when things get really bad. Ours became "This is a harsh land of many contrasts."

We heard later that the educators had arrived back in Pangnirtung two days later. There they made arrangements to visit Ross Peyton's fishing camp, where they caught lots of char and undoubtedly enjoyed having a roof over their heads and a professional cook in the kitchen. We gloated a bit because their expedition had begun with so much side, but we didn't gloat much. It had become very apparent that this harsh land, if it put its mind to it, was more than capable of sending anyone scuttling back to the comfort of a fishing camp, or making him wish he could. Hubris may be natural to some, and even invigorating, but a little of it goes a long way in this part of Baffin.

Though it didn't seem possible, the weather got worse. The temperature dropped some more and the wind rose into a 60-mph tantrum, driving the water in the air into every crevice, even through the fabric of the storm shells we wore. Walking face into it, as we did through the day, gave one the sensation of pushing into a perpetual thicket of laurel bushes. What with the wind, scrambling across moraines, wading in muskeg and fording the ice streams, fatigue began to seep into us like water into a leaky canoe, rising from the Achilles' tendons to the calves, back, shoulders and, finally, to the gunwales, so to speak, of the head.

As tiredness flowed in, warmth flowed out. After seven miles of this, we both began to get the shakes, which were not only unpleasant but surprising. Sam and I both take some pride in liking cold weather and being resistant to it. Shivering acquaintances say it is because we are both too ample; we claim it is because we are sensible feeders who maintain proper substance and circulation, and that our critics could do the same if they were not so obsessed with being skinnier than God intended our species to be. In any event, the shakes were something new for us and we decided that this was no place for metabolic macho, that what we needed fairly soon was to get much warmer and drier than we were.

This was not immediately possible, because for several miles we could not find a place flat enough for a tent that had enough shelter to keep a tent from being ripped to smithereens by the wind. Along toward mid-evening we came to a field of black rocks, an old moraine. At the edge of this field, a few feet above the Weasel River, was a shallow dry ravine. The river was choked with sheets and blocks of ice that were noisily grinding against each other. Directly across the river was a cold-looking and loud phenomenon, the 5,000-foot face of a hulking, hook-shaped peak called Thor. Snouts of glaciers poked out along its flanks, down which periodically rumbled a cannonade of ice blocks and rock splits. The names of many of the spectacular Baffin peaks are taken from the mythology of the north: Thor, Odin, Loki, Freya. This is appropriate for these brooding, violent-looking and violent-sounding outcroppings. If they are not the petrified remains of the Thunderer and his ancient associates, they are very suggestive of them.

It seemed that the ravine might do for a camp. Getting set up in such a place is a kind of teeth-gritting, play-it-one-simple-move-at-a-time exercise; unlashing soggy, frozen packs, wedging a flapping tent into the bottom of a ravine, squirming out of wet clothes into dry sleeping bags, performing contortions inside a tiny tent while trying to start a stove, preparing hot food without setting the whole delicate nest of nylon and feathers ablaze. When it was all done, it was time to settle back and start some serious worrying. Once spread out in such a situation, the spreadees are as vulnerable as uncoiled armadillos, with nothing between them and the elements except a few millimeters of fragile nylon. As the wind tore into the ravine walls and at the tent, we cringed, as much as the cramped quarters would permit, hoped we had put the thing up right, and tried, without much success, to think of what we would do if we hadn't.

Fortunately the tent held, but so did the storm, keeping us shut up in our nylon cocoon for the next day and a half. Along with worrying, we entertained ourselves by discussing backpacking, an activity about which even in good weather we have mutually sour opinions.

It is now fashionable to call deep breathing aerobic exercise; introspection has become transcendental meditation; eating clean food is organic nutrition. In the same manner, hiking has been made chic and is called backpacking. There are a number of inspirational books in which authorities explain the mystique of backpacking and how it will enable the successful practitioner to find tranquillity and truth, or perhaps God. All of which I find irritating and pretentious.

I have hiked some 5,000 miles in the past 20 years, the largest single segment being 2,000 miles along the Appalachian Trail from Georgia to Maine. I can honestly say that I have seldom taken much pleasure from the simple *act* of walking. It seems to me that the essential question is whether what you find during or at the end of the walk is rewarding enough to compensate for the hiking—or backpacking.

Considering these matters while stormbound, Sam and I decided that the best way for us to have gotten to where we were would have been in a plexiglass bubble mounted on a sturdy, well-balanced sedan chair, a modern version of conveyances enjoyed by intrepid 19th-century British explorers. So equipped, we could have seen everything we wanted in the Weasel valley and obviously been a lot more comfortable.

However, there were compensations. Midway through that uncomfortable march from Windy Lake to Thor we had paused for a rest at the top of a particularly punishing moraine. We looked back, because the wind made it painful to look ahead. Suddenly, downstream, a great white bird, a gyrfalcon, swooped down from the cliffs and took its prey, a sandpiper-type bird, on the lake flats. That was something neither of us had ever seen before but which we had both long hoped to see. Sam and I had first met, when he was 15, because of falconry. I had already been flying birds for 15 years and had written occasionally about falconry. Sam's father brought him around to hear more about its practice. Since then the two of us have admired a good many birds of prey.

Falconry is one of the most addictive sports. For 3,000 years it has held man's interest, and at times it has become a kind of mania. In medieval England, falconry was such a serious matter that a legal code evolved specifying what classes of people could own what species of bird. A holy-orders clerk was permitted only a male sparrow hawk, a piffling bird for this sport. At the other extreme, only a king could take the field with a great white arctic gyrfalcon on his fist. These pale hunters of the far north were worth king's ransoms, even men's lives.

Once I knew a young man who now has been dead a long time by reason of suicide. He was a kind of monomaniacal genius of falconry whose consuming ambition was to work with a gyr. Eventually he made arrangements to get to an area of the Arctic where gyrfalcons had been reported. He was dropped off by a bush pilot and, although he was not much of an outdoorsman, in fact not much of anything but a fanatic falconer, he found an aerie and stole a young gyr. In the course of things he became lost, was unable to find his way back to the spot where the pilot was to retrieve him, and ran out of food. When a rescue party caught up to him, he was gaunt from hunger, feverish and infected. His young gyrfalcon, however, was in good shape. He had sliced strips of flesh from his own thighs and had fed them to the bird.

Sam and I never have been falconers on the order of that tormented man, but we have some feeling for this passion. We are falconers enough that a glimpse of an imperial gyrfalcon riding down out of the sky on the back of a Baffin storm made up for a lot of backpacking.

When the storm finally broke, Sam and I crawled out of the tent and spent the afternoon drying our gear and doing a little botanizing in the moraine meadow. While so engaged we were joined by a small, elderly Korean who, under full pack, came over the rocks from the direction of Windy Lake as sprightly as a lemming. He said his name was In-Cho Chung but that his English-speaking friends called him Charley and he would be honored if he could count us among his friends. Charley, who was in his 60s, was a botanist who for 10 years had been visiting the Arctic to look at its flora, about which he planned to write a guidebook. He said that this was the second time during the brief summer that he had been to the moraine at the foot of Thor. Because a good many others had had great difficulty getting that far even once, we asked him to explain.

Flying north from Montreal, the venerable botanist had met a woman who was traveling to Auyuittuq. "I suggest to lovely young lady that we a do it together," said Charley and silenced all ribaldry with a dignified hand gesture. "We would be of aid to each other for purposes of a hiking and a making camp."

Charley and the lovely young lady reached Pangnirtung in early July in time to catch one of the last snowmobile rides up the frozen fjord. Once in Auyuittuq, things did not go well for them. "Lovely young lady was not a well prepared," explained Charley regretfully. "Very thin sleeping bag. I gave to her my feather jacket so she could stay warm, but she is a still very cold. She have trouble crossing the little streams. First I would cross with my pack. Then I would return to get lovely young lady. Very a slow."

They finally made it as far as Summit Lake, where Charley planned to spend some time looking at plants, but almost immediately the woman said she must return to catch a plane south and asked Charley to escort her as far as Pang. When they reached the fjord they found they could not travel farther by either snowmobile or canoe because of the poor ice conditions. They decided to walk the 18 miles back to Pang along the walls of the fjord. Caught between the cliffs and the ice in the water, they had an awful hike and at times had to wait until low tide permitted them to scuttle around the sheer cliffs. On one of these passages the lovely young lady fell in the water. "She was a very cold and a excited. She say 'Hy-po-ther-mia, hy-po-ther-mia,'" Charley recounted wonderingly, as if describing some curious occidental superstition. "She say will die if I do not stop and put up tent. It is a very bad place for tent, is all mud, but she say she will die so I do it."

The lovely young lady did not die and eventually made her flight. "I rest a few days in Pang," Charley said, "and now I am back. Now the plants are very nice and the weather is so good."

Sam and I met two North Americans who may have traveled farther and harder to get to Auyuittuq than anyone else there last summer. Nancy Witte and Doug Best, of Fairbanks, had left Alaska in May, traveled south to Washington state, across the U.S. to the East Coast, then north to Montreal, where they got a plane to Pangnirtung. Before leaving Fairbanks they had dried their own vegetables, jerked some moose meat and made up various whole-grain concoctions to get them through their summer of hard wandering. In Auyuittuq they hiked up through the pass, down the Owl Valley, and then back across the mountains and glaciers to emerge on another fjord where they hitched a boat ride back to Pang with a party of French mountaineers. In all, they had walked some 200 miles over the harsh Cumberland Peninsula.

Three other young Americans, Steve Amter, Rick Cronk and Ronald Sacks, all from New York City, were to become celebrities among the climbers in the pass. They were serious climbers who planned to scale Asgard, a glacier-ringed mountain near the summit of the pass that has a very large international reputation among mountaineers. Some claim that from the standpoint of esthetics and challenge it is the world's perfect peak. The three climbers waited several weeks in Pang, finally got as far as the Thor moraine but then, like everyone else, were pinned down by the storm. They began to worry that getting their gear across the glaciers to the foot of Asgard would take more time than they had and would require much too much backpacking, about which they felt more or less as Sam and I did. Occasionally during breaks in the clouds they could study the formidable west face of Thor. The face itself has never been climbed. However, peering out into the storm, the three New Yorkers thought they saw a route up which they might ascend one of the prominent shoulders of Thor, in some ways as impressive a feat as getting to the top of Asgard.

When we met them, early one evening, they had decided to try it and were loading their tubular climbing packs with enough rope, hardware, food, water and clothing for a two-day ascent. Steve, a college student and part-time Manhattan cab driver, had finished packing and was leaping about the moraine field in a kind of strange ballet. "Everything has slowed down, lying in that sleeping bag," he explained. "I've got to get my body moving before I can get my head together."

Rick said that if they made it up Thor, it would be the longest and most difficult climb he had ever done.

"Just looking at it scares the hell out of *me*," I offered comfortingly.

"I'm getting really scared," Rick said very quietly.

They cached the gear they would not need and we said we would look for it when we came back that way in five or six days. "If you come back in six days and those packs are still there, take what you want," said Ron, a tiny, frail-appearing man, but the most experienced climber of the three. "We won't need them any more."

To start their climb, they first had to cross the Weasel River, no small challenge. Stripped, they linked arms to resist the current and moved slowly ahead, using ice axes for support. They worked their way from sandbars to ice cakes, occasionally turning back and finding a new passage when the ice water became more than shoulder deep. When they reached the far beach they turned to wave. In a thoughtful mood, Sam and I went on our way toward the summit of the pass. The gist of our thinking was "better them than us."

The highest of the emergency shelters along the pass stands on a gravel bar at the edge of Summit Lake, the source of the Weasel. It serves as a regrouping place for parties that intend to climb the mountains and glaciers of the Penny ice cap, which lies northwest of the nearly always frozen lake. It was also the principal communication center for the pass. A Swiss mountaineer named Maurice, who had arrived in Auyuittuq in late June by snowmobile, had packed in sufficient supplies (undertaking two round trips from the fjord head) to make the Summit Lake camp his more or less permanent summer residence. So situated, Maurice passed along messages between parties, guarded cached supplies and in time became known as the Mayor of Summit Lake.

We left part of our gear in his care and moved up along Summit Lake in search of a nice glacier. Finding a glacier in these parts is easy. Four major ones push down almost to the shore of the lake. Getting onto one is more difficult, because the glaciers are guarded by the formidable moraines they have created. By and by, we came upon the lakeside face of a great mass of ice called the Turner Glacier and began to ascend the 1,000-foot-high guardian moraine.

The Turner moraine was very wet and slippery, being in fact a covering of large, loose rock laid not very securely on a core of ice. If stepped on in the wrong place, a great boulder the size of a Volkswagen would shift and teeter alarmingly. Below and, worse, above were equally large pieces of glacial masonry that seemed no more firmly secured. It didn't take much imagination to conceive of one suddenly being set in motion by a ptarmigan landing on a bad balance point, then rolling down the slope, pulverizing everything in its path. There is nothing flashy about moraine climbing, as there is, for example, in rappelling down a cliff, but the penalties for clumsiness are just as severe.

For the first half mile the moraine was separated from the glacier by a 50-foot-deep chasm. We worked our way up through the rocks toward a place where it looked as if the gulch pinched in and we could get to the ice. The curved turtle-back of the glacier next to us was covered with loose gravel and small rocks that lay on the ice like a rough, muddy-colored skin. Staring at it I had the feeling that I had seen something similar before or once had had similar feelings about a dissimilar thing. Shortly the memory that was nagging me came into focus.

Forty years ago, growing up in the Midwest, it was an annual custom for an adult or two of my family to take available children, early in the morning of the day the circus arrived, to the fairgrounds, where we watched the animals being unloaded from the train. Walking along the side of the Turner glacier brought back the memory of those circus mornings; specifically, of being very small and staring up at the massive, dirt-colored flank of a tethered elephant. I remember having wanted to reach out and pat the belly or rib cage of the great beast but not being able to because of adult prudence and a heavy rope which separated the pachyderms from the people. In Baffin there was something of the same feeling, an urge to pat the curved, lifelike flank of the glacier. In this case my own prudence and the crevasse were the restraints.

Eventually the crevasse narrowed and we crossed from the moraine to the glacier where its humped back was free of gravel and scree. There were snow patches on the ice, but mostly it was pure frozen water, cut with ravines and small canyons, strangely sculptured in blue, silver and lime-green ice. It was a beautiful and awesome place, but at a pedestrian's level the great virtue of glaciers, after stumbling across the moraines and muskegs of Baffin, is that they provide very good footing. The ice is level and firm, and is precarious only at the lips of crevasses.

So we strolled, enjoying the easy going, to the ridgetop of the glacier. As we reached it, the sun and a good bit of very blue sky broke through the overcast. Across the suddenly glittering ice field a 5,000-foot clock tower of rock popped into view, wisps of mist and cloud clinging to its sides like leaf mold on the cap of a new morel mushroom. This was Asgard. Whether this is the world's most perfectly shaped big rock, as some Baffin mountaineers claim, is a matter of individual taste. However, jumping out of the weather at us as it did, looming above the glaring field of ice, Asgard seemed to Sam and me a very decent mountain, well worth the trouble of finding it.

As we made our way back from the glaciers and the summit lakes, the weather became cyclical. In the mornings, from 4 a.m. to about midday, it was good, much like cool, clear, early October days in the northeastern U.S. In the afternoon, the wind rose and it became colder: clouds and precipitation blew into the pass from the fjords, and for living purposes it became February. The storms ripped and snorted until midnight and then moved seaward, and the good weather came back.

Because we intended to return from the summit slowly, looking at geological, glacial and botanical phenomena we had not seen enough of on the way up, this pattern was convenient. In the October morning we moseyed along, taking the sun and sightseeing. When clouds began to pile up, indicating that February was approaching, we looked for the lee of a nice rock, pitched the tent and waited there until the weather became more agreeable. Descending in this leisurely way, we came again, five days after we had left it, to the Thor moraine. The supplies that Steve, Rick and Ron had left, before leaving to climb Thor, were still there, untouched.

This non-happening presented us with both a practical and ethical dilemma. The probability was that the three climbers were all right; perhaps they had left the mountain by a circuitous route, had met and borrowed food from another party and were on their way back to the cache. However, having been with them when they packed and planned for the ascent, we knew that they were well past their estimated time of return and it was certainly not impossible that the delay was involuntary.

The day was turning into February and the visibility was so poor that we could see very little of Thor, let alone three small human figures. Even if we could locate them from such a distance, we had neither the skill nor equipment for a rescue. We could make a forced march to the fjord and there perhaps make contact with Pangnirtung, but this seemed like a melodramatic response and probably a pointless one, because it was unlikely that there would be anyone in Pang who could retrieve climbers from Thor. On the other hand, or third hand, doing nothing seemed like a weak choice and, should it turn out that the three were in trouble, one with which we would have to live badly for a long time.

There being no apparent good options, we turned indecisively to displacement activity—we got out the stove and started to make soup. About the time the water boiled, we heard distant shouts coming out of the mists on the far side of the Weasel. Down the moraine scrambled the three missing persons, their fists raised in the classic victory salute. They crossed the stream more rapidly and boldly than they had on the way out. When they emerged on our side, they were cold, shivering, burned by exposure, very hungry and a bit dehydrated, but absolutely triumphant. For a few moments they babbled exultantly, like players in a locker room who have just won the big game—which, in fact, the three of them had. After they calmed down, they gave us the play-by-play.

Stretching their food and water farther than they thought they could, they had made more than 25 pitches with their 150-foot rope; they had climbed almost 4,000 feet along and up the west face. It was by far the longest pure rock ascent any of the three had made, and one of the longest ever made by anyone. They had reached their objective, the unclimbed shoulder of Thor, and had come down the backside, returning through glaciers to the river. On two nights they had bivouacked in very bad weather on high ledges, but they also had had one singularly good day. "The sun was bright, almost hot," said Ron. "I was climbing in my underwear. The Weasel Valley below us from the summit to the fjord was filled with clouds. They looked solid, the color of ice. I thought this must be what everything had looked like 10,000 years ago when glaciers filled the valley. The scene was prehistoric.''

Sam and I continued slowly on down the valley, leaving the climbers to savor their triumph and gorge on food. They rested for a day and then began packing out, catching up with us again at Windy Lake. Together we walked on to the fjord head and in time found a canoe ride back to Pangnirtung.

There are mixed emotions about reentering society after such excursions. Immediately, sensually, it feels very good, as dry rooms and beds, hot water, substantial meals, a bottle of beer feel good. Also there is a sense of security. If the roof blows off Peyton Place, it is not something that must be suffered and coped with alone; coping materials and alternatives are available. It may take some waiting, but even in an outpost such as Pang you can get at most of the goods and services men have collectively devised. On the other hand, you are once again dependent upon them; on the Hudson Bay's distribution apparatus; on radio telephones; on computers that promise to make plane reservations in Montreal; on air traffic controllers in New York; on money and all that it entails; on all the arrangements and relationships, complex beyond understanding, that link each of us, like it or not, to everyone else.

It is the fact and sense of having been temporarily unlinked that is most regretfully abandoned when one returns to Pangnirtung and, inescapably, reassumes one's allotted position in the chain of civilization. That is, as the man says, the tradeoff.

They Crawl by Night

Aristotle, who so often got there furstest if not necessarily with the bestest natural history observations, called them the "intestines of the earth." Some years later my grandfather and a good many other people in southern Michigan, where I grew up, called them fishbaits, as in "Bil, slip over to the golf course after dark and get me a can of fishbaits. I want to be on the lake for the bass by sunrise."

Aristotle and my grandfather were talking about the same creature, one which through the centuries has intrigued the scientist, befriended the farmer, been the good if unwilling companion of the sport, repelled the squeamish zoology student assigned to pick apart its pickled corpse and is now the essential raw material of a growing industry. This would be *Lumbricus,* a.k.a. the earthworm.

Commonly and unthinkingly it is spoken of as the lowly earthworm. Only if used in a geographical sense (as we do polar bear or sea anemone—to associate a beast with its traditional habitat) is this a sensible name. *Lumbricus* does live below most of us. However, if we use the word lowly patronizingly, we are guilty of indefensible arrogance. Quantitatively (which is perhaps the only objective measure of biological success), earthworms outnumber us by countless billions. Qualitatively, good arguments can be made that earthworms are more important cogs in the planetary machine than are earth persons. Darwin, in addition to considering such matters as the origin of species, spent more than 40 years peering down at and thinking about the lowly earthworm. Eventually he gave us a volume with a title appropriate to such a lengthy study, *The Formation of Vegetable Mould, Through the Action of Worms, with Observations on Their Habits*. Darwin concluded that he could think of few other creatures (and he conspicuously did not think of us) "which have played so important a part in the history of the world."

Mankind in its entirety could shuffle off this mortal coil and it would be a matter of minor significance for the rest of the biotic community. In some quarters, in fact, it might be regarded as good riddance to bad rubbish. But if earthworms were to disappear, severe dislocations of most life systems would occur and complicated adjustments would have to be made—if they could be. However, not to worry. *Lumbricus* was here before we were and in all probability will outlast us, being much better dug in, so to speak. Fishbait they may be, but keep in mind who it is that is known as worm bait.

These worms belong to a large and cosmopolitan phylum of beasts called annelids, which are lumped together taxonomically by reason of being segmented, soft-bodied invertebrates. Worms squirm about from the polar caps to the depths of the oceans. Some look like flowers or feathers, some have barbs and poisons, some are sea mice and leeches; all are only a few evolutionary steps removed from spiders, centipedes and that sort of thing. Among them are some 200 members of the lumbricid, or earthworm, family: dampish, elongated, elastic beings that operate terrestrially on a transcontinental scale. They include red worms, green worms, brown-nosed worms, dung worms, slime worms, stinking worms, giant worms (an Australian species with python pretensions grows to a length of almost 10 feet) and many others most of us never will—and might not care to—meet. However, there is one we do encounter regularly and have come to think of as The Worm. This is *Lumbricus terrestris,* sometimes called the beavertail, the dew worm or, most commonly, the nightcrawler. It is the classic fishbait—although other species, notably the red worm, often serve instead.

That most of our earthworm experience and lore is based on our acquaintance with nightcrawlers is appropriate and probably fortunate. With no disrespect intended toward other annelids, the nightcrawler may be the most interesting and attractive of the lot. It is as useful, valuable, orderly, successful, pacific and dignified a wild animal as any roaming this land. "The nightcrawler is THE KING of worms," says Ray

Edwards of St. Charles, Illinois, a fishbait entrepreneur, a perceptive and indefatigable fishbait student, and probably the most enthusiastic fishbait admirer we now have. (More of the observations and opinions of Ray Edwards will follow.)

Functionally, the nightcrawler is a hollow tube of about 150 segments. One end is a head, the first 36 segments containing mouth, crop, gizzard, several hearts and, despite all the snide remarks about lowliness, not a bad brain. Tested in laboratory experiments, fishbaits learn and remember maze problems. In the field they are good at turning and manipulating a small object, say a leaf or stick, so as to fit it into their tunnels. They react to and seem to interpret vibrations, and they scent trails made by others of their kind. They have a well-developed sense of taste, preferring, for example, carrot leaves to celery but choosing celery over cabbage.

The rear 100 or so segments are mostly gut casing and, in a sense, are disposable; that is, if they are lost or damaged, they can be regrown. Up to 10 of the front-end segments can be regenerated, too. Anyone with a perverse desire to do in a nightcrawler must therefore aim for an area between the 11th and 36th segments. A fishbait is finished if crushed or severed in this vital region.

We have perhaps reached the stage of social enlightenment in which it can be publicly mentioned that earthworms are hermaphrodites. Each individual has both male and female reproductive organs. Even so, one worm needs another for mating purposes. How they do it is to snuggle up tightly against each other, heads pointed in opposite directions. Worm A passes sperm to Worm B and vice versa. The sperm is held in surface cavities along with eggs that are released and retained by each animal. A mucuslike substance is emitted from the whitish-looking ring that is prominent on all mature nightcrawlers. (The salacious commonly call this the sex band, but the proper, dignified name is clitellum.) Two worms may maintain the mating position for several hours, during which the mucus secretion surrounds sperm and eggs, then hardens into a bandlike cocoon. Eventually the worms separate. As they individually retreat into their tunnels, the cocoons are shucked off over their heads. If conditions of temperature and moisture are favorable, a baby worm will emerge from the cocoon in a month or so, when soil bacteria break down the casing. However, if the weather is bad the hatching may be delayed for a year or more. The cocoons are waterproof and well insulated.

Though these reproductive arrangements might seem complicated and bizarre, they have worked very successfully for worms. Under any moderately moist, fertile acre of ground in the temperate zone, say in front of the 8th green of a golf course in the Upper Midwest, there will be a million or so nightcrawlers. This is no more than gracious, uncongested suburban-style living from the standpoint of worms. Elsewhere, metropolises of truly impressive densities have been established. Pigpens, for example — or, more accurately, the ground under them — are for lumbricids what Orange County or Miami Beach might be to us — very desirable pieces of real estate. Under one American pigpen an interested observer found the earthworm count to be 3,388,045 per acre. Given such numbers, a precise census of the world's earthworm population is probably impossible, but within the range of the nightcrawler — northern Europe, parts of Asia, Africa, Oceania and most of North America from central Canada to the latitude of the Mason-Dixon line — there are certainly many, many billions of them.

Individual crawlers are not particularly fertile, nor do they seem preoccupied with sex the way fruit flies are. One worm normally sheds 8 or 10 cocoons a year, and from each only a single offspring emerges. However, because they do not worry about, or even know, which is male and which is female, every worm contributes to this reproductive effort and continues to do so for a long time. Fishbaits may live to be 20 years old and keep on making cocoons for most of this time — though not quite all of it. "Earthworms stop breeding sometime before they die," say C. A. Edwards and J. R. Lofty, co-authors of the *Biology of Earthworms,* a standard modern text on the subject. (The more this observation is pondered, the more astute it becomes.)

While nightcrawlers have their problems like everyone else, in many respects they are better able to cope with them than most. They are eminently edible creatures, being soft, stuffed with protein and apparently tasty, according to a scattering of human testimony as well as the responses observed in other species that occasionally get themselves a few mouthfuls. But even though a number of beasts might enjoy feasting on worms, surprisingly few are equipped to do so regularly. Shrews, moles, certain beetles and a few other diggers catch some, though not many in view of the total number of nightcrawlers. We may think of robins and other omnivorous birds as being serious worm feeders, but they only skim the surface, and there are just a few dawn and dusk hours when the light is such that worms care to surface and birds can see them there. (A note to the Great Ecologist: there seems to be a splendid opening for an owl-robin hybrid, a nocturnal worm-eating bird.)

The weather is of far more serious concern to nightcrawlers than predation, and many of their adaptations and activities are aimed at dealing with it. Water, which constitutes about 85 percent of a worm's body weight, is particularly critical. Worms do not drink as such but absorb moisture through their porous skins. To maintain their vital functions, they must operate in dampish soils where this osmotic action can occur. Being essentially thin membranes enclosing several drops of water, worms literally evaporate if they are stranded in hot, dry places. On the other hand, they can be troubled by too much water, too, because they need oxygen, which is also absorbed through the skin. They can wiggle about, breathing in their fashion, for some time in the well-aerated water of a stream or lake, but will quickly suffocate in dead or stagnant water.

Floods — worm-size floods at least — are frequent and fatal natural calamities for nightcrawlers, as anyone knows who has ever had to sweep off a tennis court covered with their slippery remains or missed a putt because a ball was deflected by corpses sprawled on a green. Most often this sort of mess occurs on a warm day after a rainy evening during which there was enough precipitation to fill worm tunnels with standing, unoxygenated water. So trapped, the nightcrawlers must get out and find high ground or drown. Once forced to abandon their familiar tunnels, they mill about in confusion. Something like trolls, who are reported to turn to stone if caught out in the sunlight, worms are usually doomed if morning finds them wandering around in the open. Many will be scoffed up by birds and other predators. Rising temperatures and drying winds will dehydrate the rest, leaving them like strands of stale spaghetti.

Nightcrawlers operate in a fairly narrow temperature range—optimally between 40 and 60 degrees. They run the risk of shriveling at higher temperatures and of freezing at lower ones. Still, extremes of heat and cold are less of a threat than the lack or excess of water, because they are at home in dirt, which is one of the best of all natural insulating materials. Like so many slippery columns of mercury in a thermometer, worms go up and down in their tunnels in response to temperature changes, surfacing on nice cool evenings and descending when the weather is bad for their purposes. During prolonged hot or cold periods, they will retreat to the bottom of burrows that may extend eight feet or so underground. There they will curl up in balls and wait for more clement conditions. If necessary, they can remain tunnel-bound, quiescent but alive, for four or five months.

The tunnel is to a nightcrawler what the hive is to a bee or a shell to a snail—an absolutely indispensable structure without which life is poor, brutal and short. Perhaps no modern authority is more emphatic on the importance of the burrow than is Ray Edwards, the earthworm guru of Maple Park, Illinois. Edwards is the author and publisher of a slim paperback, *The Nightcrawler Manual,* which so far as I am concerned is a modern natural history classic both for instruction and entertainment. A quiet, unpretentious man of 34, Edwards has evolved a unique literary technique that can perhaps be best described as writing at a shout. A passage from *The Nightcrawler Manual* demonstrates this forceful and effective prose style as it drives home some essential points about the home life of fishbaits: "Nightcrawlers DON'T live in dirt! They are TUNNEL DWELLERS. Soil is merely the BUILDING MATERIAL they use to construct the tunnel. The difference between saying 'worms live in dirt' vs. 'nightcrawlers live in a tunnel' may appear slight. Let me assure you it is of the utmost importance that you make the distinction. Tunnels are the KEYS to the survival of crawlers. *Understanding this fact is* ONE OF THE KEYS *which will lead to your success in working with them.*"

A nightcrawler is continually improving its tunnel, extending it downward—but seldom horizontally—and enlarging the diameter to accommodate increased girth as the worm grows. Considering the importance of this activity, nightcrawlers are not impressively equipped for excavation, lacking hard parts which might serve for scooping, shoveling or prying. They create tunnels by wriggling downward into crevices and moving earth from them by ingesting it. In a day they may swallow an amount equal to their own weight. This burden is then deposited on the surface as excrement, the mounds of castings which dot any wormy field. Properly speaking, worms cannot be said to eat dirt, because, except for minute quantities of nutritional minerals, dirt is taken in not for food but to provide grit for their gizzards. In this organ, just as in those of chickens and other birds, the abrasive matter serves to grind up organic foodstuffs which they collect on the surface.

Being slow, relatively weak burrowers, nightcrawlers cannot quickly construct a new tunnel if something happens to or they are removed from the old one. Thus, attempts to increase the worm population of a garden or lawn by throwing out a few handfuls of mature animals are not apt to be successful. Long before most of them can get back underground, they are taken by predators or dehydrated. Successful individuals spend their entire lives in, or at least with part of their bodies in, the same tunnel. Confusion about this situation is common and makes Ray Edwards very IMPATIENT, as he indicates in *The Nightcrawler Manual*. "Crawlers are NOT nomads. They have no way of finding the entrance to their tunnels once they leave. The worms you see CRAWLING have become separated from the only source of protection they know and are fighting a losing battle for survival. Crawling is NOT part of the NORMAL behavior of LIVING nightcrawlers. So why am I making such a 'big deal' about it? When you make an OBSERVATION of an ISOLATED EVENT, derive CAUSE and EFFECT relationships from that 'sighting,' then GENERALIZE your CONCLUSIONS, and APPLY your ASSUMPTIONS to the habits of ALL 'nightcrawlers'—*serious problems arise* which slant your thinking and fog your mind."

Edwards goes on, "It is ironic that an entire species of animals has been named in reflection of the behavior exhibited at the time they are about to die." He suggests that more descriptive names would be "night-stick-their-heads-outers" or "after-dark-comers-near-the-surface," but admits these may be a bit cumbersome for everyday use.

Though they never voluntarily leave their tunnels entirely, nightcrawlers do indeed spend a good bit of time partly extruded from them, and the acts which they perform in this position are vitally important.

Mating, for example, occurs on the surface when two neighboring worms meet by leaning out of their tunnels. They surface tail-first to deposit the casting of earth which they have swallowed underground. In feeding, the process and position are reversed. Keeping at least a few tail segments securely wedged in the mouth of a tunnel, a worm stretches out and grabs bits of grass, leaves, twigs and almost any other sort of available organic matter in its muscular mouth parts. These foodstuffs are dragged underground, where they are either eaten and digested immediately or cached in galleries. For their size, fishbaits are immensely strong. While foraging, one of them can seize and carry back to its tunnel objects that weigh 60 times more than it does. Many of the faint background rustlings and scrapings to be heard on a still summer night in grasslands are the sounds of worms moving huge loads of food across the surface and down into their tunnels.

Thus pursuing their self-interest, earthworms have a critical, continuing and immensely constructive influence on soil quality. Their tunnels loosen and aerate the earth and improve drainage. The organic matter they eat and take underground is composted more rapidly and thoroughly than would be the case if it were left on the surface. In the worm it is mixed with mineral-rich grit the animal has sucked up far underground. This instant humus is returned to and spread around the surface as castings. Each year a normal worm population will produce such topsoil at the rate of 20 to 40 tons per acre. Studies in controlled plots, where the number of worms can be experimentally increased and decreased, have demonstrated that they have a markedly beneficial effect on the growth rate, vigor and productivity of wheat, peas, beans, millet, barley, rye and meadow grass, among other floral species. In some cases the judicious addition of earthworms has doubled crop yields.

It is their function as nature's own Rototillers and fertilizer spreaders that brought earthworms to the admiring attention of Aristotle, Darwin and a good many other students. However, important as their ecological role is, worms perform it so quietly and unobtrusively that it commonly goes unnoticed and seems no more remarkable than water flowing downhill. The point where most of us meet and think specifically about worms is at the business end of a fishhook.

Because the natural range of fish and fishbaits does not ordinarily overlap, someone, sometime, must have experienced a eureka-like flash of insight into how very attractive these animals, if properly presented, would prove to bluegills, perch, bass, pike, muskellunge and even (though purists sometimes pretend it is not true) trout. When this important discovery was made, and by whom, are irretrievably lost in the debris of ancient history, but we do know that fishbaits have been fishbaits for a very long time. All the spaniels and setters, falcons and ferrets taken together probably have not made such a contribution to sport as has *Lumbricus terrestris*. To be sure, lesser members of the worm family are used, too, but the nightcrawler is the preeminent fishbait. A few segments of it can be used as a dainty morsel for small panfish; a full-bodied, vigorous nightcrawler is substantial enough to draw the attention of a bass or muskie. Despite the recent proliferation of rubber, nylon and plastic lures, neither fish nor fisherman has lost the desire for fishbaits. Last year some 600 million nightcrawlers were sold to North American anglers.

All of these worms were captured in the wild, for despite our long association with them, we have never succeeded in completely domesticating nightcrawlers. Given proper food, bedding, humidity and temperature, they can be kept in good condition for many months in a box, but true nightcrawlers will not regularly reproduce when so contained. Removal from their familiar tunnels seems to diminish their sexual appetite. This is not the case with smaller members of the family. Red worms are much less modest and fastidious and will breed willingly in bins, flats, or almost any other sort of a cage. Because of its fecundity, the red worm is the principal product of an industry that now does about $50 million worth of business a year. Most of the red worms are sold to fishermen, but there is a growing market for them among organic gardeners and pet-food manufacturers. In addition, some attempts are being made to popularize red worms for human consumption. At least one earthworm cookbook has been published, and a major vermiculturist has sponsored a recipe contest in which the winning entry was a formula for making earthworm quiche. Worm farming (or, at least, the would-be worm farmer) has also come to the attention of a variety of get-rich-in-your-basement hustlers. Complaints and legal actions against slippery worm promoters have thus far occurred in eight states. "Millions of dollars are being ripped off from the public across the country because of the flimflam in worm-growing arrangements," remarked Harvey Bell, the Commissioner of the Arkansas Securities Division, who has had trouble with hanky-panky artists of this ilk in his state.

Such shenanigans are of no concern to those who deal with true nightcrawlers, because the nature of the animal restricts them to the role of hunters and trappers. Edwards, in his invaluable *Nightcrawler Manual,* deals bluntly and instructively with this matter. "I have repeatedly seen articles about red worms (sometimes called hybrids or wigglers) which are accompanied by photographs of some guy holding up a 10-inch nightcrawler. These methods [employed in red worm culture] CAN NOT be used successfully with nightcrawlers—a fact which they neglect to mention in the story! IT IS IMPOSSIBLE TO RAISE NIGHTCRAWLERS IN YOUR BASEMENT OR GARAGE."

With the help of crews of high school students whom he trains, Edwards himself collects some 500,000 nightcrawlers a year in the vicinity of Maple Park. They are stockpiled in holding pens at his home and are sold directly from there or from vending machines which he has adapted to dispense worms, and which he has placed at a number of fishing sites in northern Illinois.

Edwards, once a junior high science teacher, is now a full-time fishbait dealer. "I got started," he says, "because I was a bass fisherman and I was cheap. I was too cheap to keep paying out 60¢ for a dozen nightcrawlers, so I started catching my own. Then I got interested in making money and trying to build up a business. I also got interested in the nightcrawlers themselves, watching them, reading about them, trying to figure out their habits. I started out being a fisherman, but now I never seem to have much time to fish. The truth is that on a good night I maybe find it more enjoyable to go out hunting worms and thinking about them than I would if I went fishing."

Having become something of a tycoon and a considerable fishbait naturalist, Edwards spends much of his time selling worms, manufacturing and serving his vermi-vending machines, lecturing to various sporting groups and writing about worms. More and more he depends on his cadre of teen-age associates for the actual collecting, but he remains the grand master. There is probably no other American who has done more to elevate worm hunting to the level of a science, or an art, or a sport. As a great fan of *The Nightcrawler Manual,* I have come to believe that Edwards is to fishbaits what Izaak Walton or Sparse Grey Hackle is to fish. Therefore, I called him and asked if I could follow along sometime when he went worming. Being very generous of his time when it comes to promoting serious interest in fishbaits, Edwards agreed.

We met one warm summer evening at a café some 40 miles west of O'Hare Airport. Instruction began immediately. Edwards allowed that there was no need to rush off because it is foolish to try to stalk the wily *Lumbricus* until it is fully dark, a condition which at this time of year in the Upper Midwest would not occur until almost ten o'clock. Therefore, we sat around drinking coffee, eating a wretched, prefabricated, frozen Key lime pie and talking about worms. Edwards was apologetic about the weather, saying it was too dry for really good hunting; he was hoping that there would be enough dew so we could get a little action. He wished I had been able to come several days earlier when a midday shower had created ideal conditions in the evening. "I was out with my crew and we picked 20,000 or so," he said. "We were getting some real ropes." (Exceptionally long, fat animals are ropes, or snakes, in the jargon of worm hunters.) "The next morning when I was packaging them, I dug out a double handful of about the best-looking crawlers I have ever seen. They were like prize tomatoes or squash you'd show at a fair. I took them in to show my wife."

"Did she like them?"

"She said to get those worms out of the kitchen."

Edwards said that we would be hunting on a golf course and was a bit defensive about selecting such a site. This was surprising, because in the days of my youth golf courses were regarded as absolutely the best places to look for worms. Edwards politely explained that my lore, like that of many others of my generation, was obsolete. Because of continued applications of herbicides and pesticides, modern golf courses are often worm-poor. "Around here we get our best picking in vacant lots, on school grounds, and around ball fields where they don't use many chemicals." (Ecological aside: investigators have found that earthworms can tolerate relatively large amounts of many commercial pesticides, residues of which are often found in their tissues. Unless the doses are massive, the worms continue to function without immediately apparent ill effects. However, because the toxins tend to be concentrated in their bodies, other less resistant creatures, such as birds, may suffer more from eating contaminated worms than they would from direct exposure to the poison.)

Despite the drawbacks, Edwards had chosen a golf course for demonstration purposes, thinking it would be a more scenic and convenient place to take a rookie than some of the scruffier but more productive areas where he and his commercial crews hunt. "This course isn't the best place, but it's not bad," he said. "I guess it's a little run-down for golf. They don't seem to spend a lot of money taking care of it, but that's good for worms. Also, it's a lot easier than crawling around in broken glass, which we have in some of our spots."

We drove to the golf course, and there Edwards began to get into and explain his worm-hunting costume, an outfit he has created as a result of his experiences in the field. The basic uniform is coveralls and kneepads, worn for protection and convenience while crawling around on the ground in the dark. On a belt around his waist he hangs half a dozen plastic boxes — gallon milk containers with the tops cut off. Each of these will hold 500 or so worms. In a similar container he carries a few pounds of sawdust, which serves the same purpose resin does for a baseball batter. "Always keep sawdust on your hands," Edwards said. "If you don't, they [the fishbaits] are going to slip out or you will squeeze too hard trying to get a grip. You get dead worms." (Oak and redwood sawdust should not be used for this purpose, their acid content being harmful to worms.)

Edwards has some strong opinions about footwear, and in *The Nightcrawler Manual* warns: "NEVER WEAR heavy boots. Any heavy, clunky, hard-soled footwear makes too much 'noise' as you walk in the grass." Tennis shoes are good enough for casual amateur hunters, but he himself is more professionally shod. What he pulls on is a disreputable-looking pair of old, cracked dress shoes from which the heels and stiff outer soles have been removed. "The inner sole that remains," he writes of his innovation, "will be soft and thin. As you walk slowly and silently at night you will actually 'feel' worms popping under your feet when you step on them."

A critical piece of equipment is the worm light, a battery-powered lamp worn on a headband as miners or cavers do and equipped with a rheostat switch so that the degree of illumination can be adjusted. Bright, clear light alarms crawlers and drives them underground, so Edwards covers his headlamps with white plastic to soften and diffuse the beam. ("Note—I didn't say clear plastic, green plastic, yellow plastic—I said, and meant, white plastic.")

Some authorities have recommended using red headlights, but Edwards believes this is Unnecessary and Unwise. "The worms don't care whether the light is red or white so long as it is soft. The difference is that it is harder for you to see them using a red light. You can't pick what you can't see. The trick is to keep adjusting your white light so you get enough for good picking but not enough to scare them. It's a little different every night, depending on the sky light."

So prepared, we set off across the ninth fairway, Edwards pussyfooting along in the lead, waiting for worms to pop under his stripped-down shoes. After a few moments he turned down his light and called back that he had found a hot spot. Considering how surreptitiously we had been advancing, he spoke very loudly and clearly. "They don't hear like we do," he explained. "You can talk and even yell and they don't pay any attention. If I'm alone and there's a ball game or something I want to hear, I carry a transistor radio. The thing they are sensitive to is vibrations. Very sensitive. If you walk heavy-footed, you squish the ground. That scares them and they go down."

Trying to squish as little as possible, I joined Edwards at the hot spot. A number of nightcrawlers were visible, eating, depositing castings, loving or perhaps just taking the night air. We got down on all fours and had at them. For those who may have scooped up a few sick, sluggish animals stranded on a sidewalk, or plucked an occasional one from a lawn, it might seem that hunting nightcrawlers at professional speed would be simple, dull stoop labor on the order of harvesting stringbeans. IT IS NOT. They are slippery, very alert and agile creatures. One false move or vibration and they disappear into their tunnels with the speed of a coiling watch spring. During the course of two or three minutes, Edwards, alternating hands in an easy plucking motion, caught 50 or so worms. In the same period, favoring a lunge-lurch style, I got eight. Two of them were probably squeezed too hard and another had been indisputably severed in the neighborhood of segment 54. Edwards was kind enough to say that I was not as awkward as *some* beginners he had seen, and then proceeded to offer advice aimed at improving my technique.

Long experience has enabled Edwards to break down the act of catching a nightcrawler into its step-by-step components, and for instructional purposes he explains them as Ted Williams might describe what to do about the low, inside curveball. As in so many other physical activities, the initial stance is of cardinal importance, and Edwards has developed and recommends something he calls the Elephant Walk. To demonstrate, he assumed a crouching position somewhat like that of a down lineman. Spreading his knees apart as far as possible, he crawled ahead while weaving his head and shoulders from side to side in a manner suggestive of a trunkless elephant. As he proceeded, he gleaned worms on both sides, picking with one hand and balancing with the other, then reversing the hand positions. Edwards claims that the Elephant Walk enables the hunter to spread his weight so as to reduce the vibrations he makes and allows him to reach the maximum number of worms with a minimum amount of movement.

Having thus sneaked up on a colony of surfaced worms, one's next movement is the grab, which must be made with controlled speed. Edwards says that in a pre-grab position, the hand should be open, the thumb extended in a hitchhiking gesture. It is then closed around the worm so that some of the head of the beast extends between the thumb and index finger and as much of its body as possible is pressed against the palm. Seizing a fishbait between the fingertips is amateurish and ineffective, because not enough hand-worm contact is provided. As a training aid, Edwards suggests that students boil themselves some rice, line up the grains to simulate a mature fishbait and then practice grabbing until the technique is mastered.

Finally, there is the pull. Steady but gentle vertical pressure should be applied to the grabbed worm. Muscular as they are, a fishbait whose tail is well wedged in its tunnel can resist removal like a tiny mule and can be yanked or jerked apart before releasing its grip. Edwards points out that if a worm has been correctly seized he cannot escape, and a good hunter will patiently maintain tension, just as one might with a hooked fish. In time the worm will tire and can be landed, so to speak. With very strong, stubborn animals, another trick may be used. If the grab has been made properly, the head of the worm should protrude through the fist. The hunter gently squeezes this with his free hand, and often the beast will relax its tail grip on the tunnel wall. Edwards thinks the reason is that pressure on the head parts cons the worm into thinking it is in the safety of its burrow.

Perhaps because of the relatively high temperature, but more likely because I was a noisy elephant crawler, a clumsy grabber and erratic puller, the hot spot we were working petered out shortly, the nightcrawlers retreating underground. However, because this was essentially a sporting and educational outing, Edwards was not especially concerned and used the situation to demonstrate another bit of worm craft. Reaching into his collecting box, he took out a nice handful of fishbaits and scattered them about the ground as if broadcasting grass seed. "We can pick these guys up anytime," he explained, "but until we do, they are going to crawl around like crazy. They are desperate because they aren't in their tunnels. The ones below are going to *hear* them, as we would say, but actually they are going to feel them banging around on top. Now what are they going to think? They are going to think there is no danger, because these guys are moving on the surface. The ones below are going to think the ones on top are having a good time, probably feeding or mating. So they are going to want to come up and get into the action themselves. This is just like fishing or any kind of hunting. You have to try to put yourself in the position of whatever it is you are after."

In this case the ploy worked very well. Within 10 minutes or so, more worms than were originally in the spot began poking their snouts out of the ground. When a worthwhile number emerged, they, as well as the decoy animals, were grabbed and pulled.

"A lot of beginners make the mistake of thinking that once they have picked a few minutes in a hot spot they have to find another, because the worms have been scared away," said Edwards. "They don't understand how worms live. A nightcrawler can't run away. It has to stay in a tunnel, and there is only one opening on the surface. If you see it once, you will find him there again in exactly the same spot. Sometimes when I first start picking in a spot, I'll see a real rope just out of reach. It's just like hunting. There is a challenge to get that big one. I'll mark the spot where I saw it go down with a little sawdust and come back later that night or even another night and get it."

"What's the biggest one you ever caught?"

"It was a couple of years ago. I saw this really unbelievable worm, the snake of all snakes, and you can bet I hung around until I pulled him. He was as long as my lower arm from the elbow to fingertip and about as big around as my thumb."

"What did you do with him? There is no way you could stuff him in a vending machine."

"I decided since it was the biggest worm I'd ever seen, I was going to try with him for the biggest fish I'd ever catch. There is a lake 10 miles or so away that not many people know about, and it has some lunker bass. I took that rope out there, strung him up and put him in the water."

"What happened?"

"I caught a six-pound carp. What a disappointment."

"You should have stuffed that worm and hung it over your mantel."

"I thought about that afterward."

Trailing

I grew up in a family of hunters, trappers, poachers, outdoorsmen, agricultural workers and vocational and avocational naturalists, among whom reading animal sign was as common as reading racing forms may be in other clans. Even today, the pastime of trailing—the finding, following and interpreting of tracks and other signs that animals leave behind—remains a favorite of mine.

Trailing is not a particularly uncommon interest, even though it is seldom given a formal name, and no clubs, catalogues or industries have spun off from it. Trailing is something like throwing a ball. Some are more into it than others, but almost everyone has had some experience with it. I have yet to meet anyone who, upon accidentally discovering cottontail tracks on a suburban lawn or a muskrat run at the edge of a water hole on a golf course, is bored by the discovery. The deep-rooted appeal of trailing may lie in that it provides an outlet for our hunting instincts even though it does not necessarily involve the scragging of beasts.

Like any sport worthy of the name, trailing is physical to whatever degree one chooses. It can be no more exerting than playing golf; it can also be decidedly demanding. It is sensually stimulating. Eyes, ears and even noses must be alert. Mental activity, both inductive and deductive, is required, not only to find signs but also to try to figure out what they mean. Good trailing is something like reading a good mystery.

Some places are better than others for trailing, although one can get in a little practice almost anywhere. A few years ago when I was on my way to the Yukon to look for, among other things, wolves, I planned very badly and had to stay overnight in New York City. That night Baghdad-on-the-Subway and its airports were closed down by a few inches of snow, and the next morning, waiting for things to get moving again, I occupied my time by following stray alley cats along and behind buildings between Riverside Drive and Broadway. Because I like the Yukon valley better than West 80th Street, I found cat-trailing inferior to pursuing wolves, but it was nonetheless genuine trailing, and a better way to kill a morning than hanging around Zabar's.

I like trailing enough to have gone back to the Huachuca Mountains of southeastern Arizona half a dozen times to play at it. The Huachucas are one of a series of formations called "mountain islands" because of the way they rise precipitously out of the Sonoran Desert. They are disconnected ranges, with peaks up to almost 10,000 feet, that extend from the U.S.-Mexico boundary 50 miles northward. Along with some lesser islets are four major massifs, the Huachucas, the Chiricahuas, the Santa Ritas and the Baboquivaris, which lie roughly parallel to one another, separated by wide, grassy valleys.

These ranges used to be hunting grounds and, in the time of Cochise and Geronimo, hideouts for Apaches, Yaquis and Papagos, although permanent tribal settlements were small and scarce. The mountains are still sparsely populated. Spruce and pine that grow near the summits were logged by homesteaders, and there has always been a little prospecting and mining for copper, gold, silver, zinc and other ores.

Virtually all the mountain land is now held by one federal agency or another, the principal one being the U.S. Forest Service, which grants leases to ranchers to run cattle through the highlands. This does not bar other uses of these leased public lands, and they are open for trail riding, hiking, hunting, camping, nature loving. Fortunately, either by design or lack of funds, the Forest Service has not done much to improve the mountain country for recreation. The roads into it are few and often impassable, and permanent facilities are meager. Because there has never been much pressure to make them otherwise, these ranges remain wilderness islands, standing above the golden-age retirement villages, the get-rich-with-pistachio-nuts ranchettes and the other Sunland real estate ventures that sprawl south across the desert from the Phoenix-Tucson urban conglomerate.

While other resources in the mountain islands may not be exceptional, the native flora and fauna are. In terms of diversity of species, this small, sharply defined highland area is unique in the U.S. and rivaled by only a few other locales in the world. The Huachuca range, which I know best by reason of having spent a year in it studying coatimundis, tropical members of the raccoon family, is only about 30 miles long and not 10 miles wide at its broadest point. Yet in these 200 or so square miles, ornithologists have identified breeding birds that represent 25 percent of *all* the species that reproduce anywhere in the continental U.S. There are golden eagles and coppery trogons, roadrunners and ravens, 14 species of hummingbirds and occasional parrots.

The same situation prevails with mammals, some 70 species of which—more than inhabit the entire state of Illinois—have been found on this single mountain range or in its immediate environs. For example, all three members of the raccoon clan found in the U.S. are residents of the Huachucas. The raccoon "proper" and the ring-tailed cat are widely distributed, but the coati—an abundant animal in South and Central America—is seldom found in the United States except around the Huachucas and the other mountain islands of Southern Arizona.

Wild hooved stock is represented by the whitetail and mule deer, the javelina or peccary and the antelope, and within living memory there were big-horned sheep. The grizzly has been shot and poisoned out, but black bear remain. The coyote, the gray fox and the kit fox are found regularly, and now and then a lobo drifts north from Mexico. Mountain lions and bobcats are resident. The jaguarundi, a slim, water-loving tropical cat, is a very occasional visitor. There are even rarer but verified reports of the jaguar and rumors of the ocelot. There are 16 species of bats, four of skunks, a similar number of rabbits—including both desert and eastern cottontails—three of squirrels, one each of badger, porcupine and opossum and a great array of shrews, rats and mice. Plant, insect and reptile communities are similarly cosmopolitan.

This remarkable variety reflects special environmental circumstances. The fundamental one is that in the U.S. there is no area so high so far south, and therefore many diverse climate zones are jammed together in a very small space. In deep arroyos at the foot of the mountains, conditions are subtropical. Above are desert, short grass prairie and oak, juniper, sycamore, walnut, pine and fir forests. At the top of the mountains are stands of aspen and alpine plant communities.

The peaks, rising so sharply from the desert, intercept weather systems and receive almost twice the precipitation that falls elsewhere in south and central Arizona. When it is bone dry in the rest of this arid country, there will be mists and showers on the highest ridges. Above 8,500 feet, snow persists from December into May.

Water dribbles out of springs, runs down the slopes, dampens the canyon bottoms and the valley grasslands and even collects in bits of bog. Everywhere the wetness combines with the elevation to further diversify the habitat. In a 10-mile walk across the Huachucas one encounters microenvironments that approximate nearly all the macroenvironments between Mexico and Canada.

The extraordinary ecology encourages trailing, which I have enjoyed in many places but never as much as in the Huachucas. The best time to be there—the best time to trail anywhere—is winter. Snow cover or, at least, moist muddy ground makes a better surface for this sport than bare rock and hard, baked caliche. Last winter local operatives called me shortly after the first of the year to say that there was an unusual amount of snow, down to the 4,500-foot level, that the canyons were filled with water and that the ground was splendidly plastic. In response, as others headed toward Maui and Martinique for their midwinter R&R, I set off to the mountain islands for mine.

Montezuma Canyon is the southernmost major canyon in the Huachucas, and I make a beeline for it whenever I am in southern Arizona. In an ascending order of importance, the reasons for its attraction are:

One of the few roads leading into the mountains suitable for such transport as a rented car goes toward Montezuma.

An important part of my heart remains in this canyon. About 6,000 feet up, half a mile from the road, is an old stone cabin where three students and I lived for a year, studying a tribe of coatis then occupying Montezuma.

Will and Deane Sparks and their brood of little Sparkses live there.

They were our nearest neighbors during the coati year and remain our close friends.

I first met Will Sparks when he was remodeling a mining camp that had been in his family for generations. He is a large man who looks and acts remarkably like John Wayne when the Duke was in his prime. Will can strap on a pound of turquoise or a .45, sit or shoe a horse, play a soft guitar around a campfire or dance like a butterfly-bear to one, keep a bar in laughter with country stories or empty one in righteous indignation—all without pretense. There is not a bit of rhinestone about him. By birthright, experience and temperament he is a natural Western man; a compassionate conservative, a libertarian populist, a tender hardhead. Especially, he is a Huachuca mountain man, having roamed this island since his boyhood without ever having lost his admiration for or curiosity about it.

I had first visited the Sparkses' place when we were scouting the mountains, trying to find some evidence of coatis. Will and his brother-in-law Bob were knee-deep in mud, working on a well. I asked them if they knew anything about the animals. In the slow, choreographed Western manner that lends emphasis to words, Will put down his tools, leaned back against a rock, squinted and said, "Yes."

(We talked recently about that first meeting. "I figured you were a smart Yankee," Will said, "who was probably surprised a dumb redneck would know a chulo bear from a prairie dog." Chulo is the border name for the coati.)

Had they seen any chulos recently? Will did another slow study, with Wayne body English. "About 10 minutes ago, if you count that recent," he said.

For us, it was as if somebody had said yes, indeed, I saw a little fellow with a funny hat burying a pot at the end of that rainbow over yonder. The coati study had begun as a gamble. Though these mountains are one of the few places where they are ever found in this country, they are not always present. Tribes of the animals move north from Mexico, stay a while and then either return south or die out. When we started, the considered opinion of game wardens, forest rangers and field biologists was that the coati population in the Huachucas was very low or nonexistent. We were very lucky, having arrived at the beginning of an inexplicable up cycle. The animal hanging around the Sparkses' mine was a male, a member of a tribe of 25 coatis that we were to find later and live close to in the upper part of Montezuma Canyon. That morning all of this was unknown, but Will's report was the best one I had heard and I had a lot of questions that I began asking.

"Probably the best thing would be to go look at that rascal," Will said after I had carried on some. "He ought to still be there in a tree if," and he fixed his brother-in-law with a deadpan stare, "he isn't dead from eating those biscuits Bobby baked this morning."

The chulo was draped over the limb of a sycamore in a somnolent state. I gawked for a while and then spent the rest of the morning talking to Will, both of us making the surprising discovery that despite a lot of obvious differences we were very compatible. In the course of conversation, Will said that he thought chulos were the perfect animal: "They take good care of their little ones, and they don't bother anything that doesn't bother them. But if it comes to trouble there isn't anything in 700 miles that is going to have much luck tangling with a tribe. I've seen lions back off from chulos."

That opinion alone recommended Sparks as a sound man. Everything since has confirmed my first impression. In fact, so far as I know, he has only one serious character defect. He is what is known locally as a "workin' sumbitch," and it is hard to drag him off into idle pursuits of the sort that are my specialty. The mine (zinc) is no longer economically viable, and Will now owns an earth-moving and construction business. He operates under the theory that no job will be done exactly right unless he does it all himself. Because of this, he has a good business but one that usually occupies him 12 or 14 hours a day. Last winter, fortunately for me, Will decided God would probably not strike him dead if he took a day off and went trailing in the mountains.

"Good," his wife said. "I've never understood why you should go through all the grief of working for yourself if you can't do what you want when you want sometimes."

We took Will's best four-wheel-drive truck to see if we could get to Copper Glance, a canyon 10 miles to the north on the western side of the mountain. There are times in midwinter along this border when the weather compares favorably with anything any poet has seen in June or

October anywhere. We hit one of those days. A storm front that had been dropping snow showers for the previous 48 hours had passed on to the northeast. It left behind skies so blue they seemed almost black in contrast to the few wisps of clouds that hung around the peaks. The air is usually dry and refreshing in this high country, but on a poststorm day it seems to crackle in the lungs. The sun was hot enough so that we could walk through the snow in shirtsleeves. The snow, pure white and soft, was so dry and powdery that if you picked up a handful and squeezed it, it spurted out of the fist like fine flour or dust. A few inches of this delicate precipitate lay on the ground above elevations of 5,000 feet.

The rutted road we were traveling is the only public one along the western beach of the mountain island. It overlooks the San Rafael valley, and there are 30-mile vistas south into Mexico and west to Santa Rita mountain. As we were slithering along we saw another truck coming toward us from the rolling grasslands below. "That's CR," Will said and stopped to wait. Especially in the winter, it is rare to meet anyone on the west side of the mountain, and people who are not strangers invariably stop to pass the time of day. CR was a bull rider on the professional rodeo circuit and is now foreman on a ranch in the San Rafael valley. In addition to saying howdy, CR was interested in a piece of four-inch steel pipe that Will had shaped and fixed to the truck as a rear bumper and tow bar.

"I really got to get me one of those. That is a slick job," CR said several times and then asked what we were up to. We said we were thinking about leaving the truck and walking up Copper Glance just to see what was happening there. CR was almost as astonished as if we had said we were going to try to swim to the top of the Huachucas. This is riding country. People used to ride horses across it, and now they mainly ride pickups and ORVs. Beyond what it takes to get to a corral or a garage, very little walking is done at all, and there is no tradition of walking for pleasure. When my student researchers and I first came to the mountains and began hiking around them looking for coatis, it raised doubts about our sanity and trustworthiness, because people on foot are apt to be illegal aliens, radicals, environmentalists or worse. After we established ourselves as harmless eccentrics, we received well-intentioned warnings that if we kept on walking we would probably be struck down by rattlesnakes, lions or heart attacks. In time, we earned a certain amount of respect, though no converts. We began getting inquiries about what things were like in odd corners of the island that people had seen from a distance for years but had never investigated because there was no way to ride into them.

CR queried us closely, shuffled and spit, took a deep breath and said, "You know, this is such a pretty day I'd almost like to get out and walk up there with you if my knees weren't so bad."

I had been in the Huachucas only two days this trip and had already heard two stories about CR's knees. Will's was the better. CR had banged them up badly while trying to ride bulls on the rodeo circuit. He had hurt them again wrestling with a cow on the ranch. In the old days he probably would have spent the rest of his life a half-cripple, as so many cowboys have for the same reason. However, the Eastern magnate for whom CR works is of a different culture, a man of different means. He arranged for CR to fly to Boston for a complicated restorative operation. According to Will, when CR checked into the airline counter in Tucson he was asked if he wanted a seat in the smoking or non-smoking section. "CR told them he didn't want either." Will said. "He wanted the chewing and spitting section. They said they didn't have a chewing section. He said that was all right, just put him by a window."

"It was one of those little fellows in a fancy uniform," CR said, confirming the story. "It really did shake him up."

CR wished us well and rode off into the sunrise. We turned down a Forest Service road, slid across a few cattle guards and finally left the truck in a grove of Emory oaks on the flats a bit to the north of the mouth of Copper Glance. There were coyote tracks in the thicket. At least one animal had been there only a few hours before because the grains of snow thrown up around the prints were still distinct and fluffy, not melted by the sun. The coyote had been nosing under logs and pawing at dry clumps of yucca that are often refuges for mice. This animal had probably been having a dull time of it; rather than carrying his tail curled alertly over his back, he had often dragged it absentmindedly.

Another dog—a fox—had been making somewhat the same rounds and had occasionally used the trail that the coyote had broken in the snow. I thought it was probably a gray, simply because they are very common, but Will wasn't convinced. He thought it might have been a kit fox because the prints were very small. "They've come back since you left," he said. "I saw two or three this fall."

"We saw only one the whole year we were here," I said.

"I love to watch those little fellows," Will said. "The way they tippy-toe along like they were walking on eggs."

Besides tracks there are a good many other signs that contribute to trailing—displaced rocks, broken vegetation, bits of hair, tunnels, holes. A dependable one is scat, which every mortal being must sooner or later leave behind. Scats provide a somewhat more complete record than tracks, indicating not only that a creature was in a given spot at a given time, but also something about what it had been doing at least a few hours previously. A raccoon dropping that contains crayfish shell and is found high on a dry, rocky ledge is evidence that the animal has ranged at least as far downhill as the nearest stream. Most mammalogist camps have a scat bag, and there are always scats to be dissected and recorded. It is tedious, though not particularly repulsive, work, and it is a good source of information about feeding habits, foraging territories and ecological relationships.

Scatology suggested that the winter had been a fairly hard one, at least for the coyotes and foxes that had been hunting the oak flat. There was scarcely any hair or other animal remains in the droppings. They were made up almost exclusively of traces of juniper berries, which are the C rations of these mountains, the food everything from rodents to carnivores turns to when other pickings are slim.

The first creature we met in the flesh was a roadrunner. There is a certain accuracy about the cartoon characterizations of roadrunners. They are indeed birds with an irascible, arrogant, vaguely rascally air. This one jumped out of a patch of agave, gave us a belligerent, go-to-hell glare and then made one of its patented sprints, disappearing in a ravine.

"I was in Ash Canyon in summer," Will said. "We hadn't had any rain in months and it was so dry you could hear the centipedes rustling in the manzanita leaves. Suddenly there was a terrrible commotion going on up above me, and when I got there I found this roadrunner flopping around in the middle of a patch of beaten-down weeds. Both its wings were broke and probably other things. The sorriest coyote you ever saw was just lying there a few feet away. Its eyes were gone, and he had blood all over him from a lot of gashes. I shot both of the poor things, but that bird had put up some kind of fight. If it had been a man we would've pinned a medal on him and made him champion of the world.''

We forded a stream and for the rest of the day climbed along it. Because of this watercourse, Copper Glance is one of the places in the Huachucas where there are small, wet meadows. On one of them, in September after a rainy spell, I had an experience that would be a good one to remember on the day I die. My son Ky and I had spent the morning working slowly up Copper Glance. We stopped to eat lunch in a meadow that lay in the full sun. The meadow was carpeted with wild flowers, and the flowers, the grass, the bushes and everything else were coated with butterflies: swallowtails, sulphurs, skippers, satyrs and others that were exotic to us. Ky and I wondered how many butterflies there might have been in that meadow—thousands certainly and probably tens of thousands. For them, we were simply two additional perches. As we ate we gently brushed them away so as not to crunch up any wings, but when we were finished we lay back and let them alight. For 15 minutes we lay under spangled blankets, brushed and delicately massaged by the butterflies.

There were no butterflies in Copper Glance on this January morning, but the aesthetics were almost as remarkable. In the tight canyon it was windless. The snow lay much as it had fallen the previous day, in puffy carpets and mounds on the ground and in free-form patterns on the needles of the pines and firs and leaves of the live oaks. We came into the canyon about the same time the sun did, and it backlit the woods and the ice formations along the creek, making them glimmer like crystal chandeliers. We came upon six white-tailed deer, which stood motionless for an instant under a huge, shining alligator juniper, as if posing for a Christmas card or a Disney producer.

Will is a backslapper or, more accurately, given his size and exuberance, a back-smasher. He beat on mine and said, ''A man could have a million dollars and not have what we've got right this minute.''

After the storm had passed, most of the inhabitants of the canyon had come out frisky and hungry from whatever shelter they had taken. Fresh tracks were everywhere, almost too numerous for our purposes. Deer, squirrel and rabbits were so active that their sign tended to obliterate everything else. A ring-tailed cat, the smallest and shyest of the native raccoons, had been working in a pine grove through which we walked, but we wouldn't have known had we not noticed a faint trail diverging from a hodgepodge of cottontail signs and heading toward a cliff where rabbits would have no legitimate business. We followed far enough to find tail-drag marks and a decent set of prints made by the darting, light-footed animal.

In a shallow draw a number of deer had churned up the snow, but they had missed a patch on the lee of a boulder that held the impression of a single, roundish paw. There had to be others, and we cast around, circling the deer mess until we picked up more of these prints. We stopped to have a smoke and figure out what we had found. In theory and fiction, trailers immediately recognize every sign they see, but my experience has been that this is not the case in practice. Sheer ignorance aside (nobody is apt to know all the signs he comes across), the evidence is often obscure and the possibilities various. The general procedure is to apply Ockham's Razor: test the simplest possibility first, and if that does not prove out move on to others. Our deliberations and conclusions were more or less as follows.

In these parts the most commonly found, largish four-toe-and-pad tracks are likely to have been made by one of the canines. However, the first clear impression we found eliminated this possibility. There were no claw marks at the end of the toes, as there would have been had a canine left the sign. This was a cat of some sort; cats usually retract their claws while traveling. There were two reasonable feline possibilities: a mountain lion or a bobcat. The lion is a much larger animal, but in this case the impression had spread and softened in the dry snow and size alone could not settle the matter. Either a small lion or a large bobcat could have made the track. We began to speculate about behavior. The trail led upward onto the canyon wall and disappeared in a jumble of boulders and bare ledges. Either of the cats could have gone that way, but lions are a bit more inclined to follow open established trails. Also, we found no tail marks, which a lion can—but does not always—leave and a bobcat cannot. Will made another point: "If it was a lion that small, it probably would be this year's. But mostly when they're that young they're still traveling with their mamas, and we only have one here."

The circumstantial evidence suggested bobcat, but there remained one other remote possibility. It was impossible to absolutely eliminate the ocelot. There had been no record of one in the mountains in years, but they had been there in the past and one could still stray in—though if one did, it was very likely nobody would know about it. The strong probability was that a bobcat had crossed Copper Glance canyon during the previous evening; the ghost of a chance that it had been something else is why the Huachucas are such an interesting place to trail.

"Did you ever see one?" Will asked about bobcats.

"Not here, but twice in the East, just for a second or so," I said.

"I saw one here, in Bear Canyon. It was gone like that. They're the hardest thing there is to see. Look at us, how much we've been out trailing and we've only had about 15 seconds' worth of bobcat between us."

Beyond the fact that in both activities a lot of time is spent walking along head down looking for something, there is another similarity between golf and trailing. When you hit a golf ball, it is generally aimed at the hole and there is a possibility it will go in, but the chance is remote and is seldom realistically anticipated. In the same way, signs can be followed until their maker is found and run into its hole, so to speak, but this involves so much luck and work that it is not a practical reason to go trailing. Yet as a climax to this splendid day we brought off the equivalent of a hole-in-one, or at least of sinking a long approach shot.

Copper Glance is a good place for javelina, the tough, pugnacious little wild pig that roams the Southwest in small family groups. We may well have passed some of their hoofprints on the way up without identifying them because of the erasures and smudges made by the deer. At about 7,000 feet we came on a very straight trail made by a number of somethings that had crossed the stream going to or from the wall of the canyon. Considerable recent use had turned the trail into a mush in which no individual prints were clear, but the probability was that they were javelinas because they move along in close-order groups, rather than cavorting about individually like deer. Backtracking along the trail confirmed this guess. It led into the mouth of an abandoned mine tunnel. No deer would go into such a place, but javelinas very obviously had: tracks aside, their scent, a strong, skunky one, was evident. Cautiously, with an eye to quick escape routes, we tossed a few pebbles to see if we could flush out anything.

Javelinas have a local reputation for ferociously attacking anyone who gets in their way. I met one hunter who claimed great familiarity with these savage beasts. He said they would chase a man up a tree and then wait in a circle below it until he starved to death, fell out and became pig food. The man was a Phoenix road hunter who tended to be terrified by things both real and mythic in the mountains. Yet there is a basis of fact in the javelina attack stories. They are shortsighted and upon hearing or smelling something strange often will trot closer to see what it is. They may come directly at you for this purpose, but they are easily avoided simply by stepping out of their way. Generally, it is prudent to do so. They are not savage killers, but for a 40-pound animal standing no higher than a boxer dog they have extremely powerful jaws with formidable teeth, two of which are tusks. Now and then somebody ends up with a badly mauled leg because of getting crossways with a javelina.

There was no response to the stones we threw in the tunnel, so we went back across the stream and, heads down, began to follow the javelina path in the other direction. Our heads came up quickly only a hundred feet or so from where we started when we heard a sharp, commanding grunt. Ahead of us stood an obviously agitated sow. Her bristly mane was raised, she was pawing the ground audibly, and she was suggestively gnashing her teeth. Farther up the slope, three animals that looked to be yearlings and a mature boar were unconcernedly uprooting and munching away on prickly pears. We stopped dead in our tracks and soon saw why the sow was so exercised. Two newly born offspring (an umbilical cord was still attached to one) stood just behind her in a rough circle of bare pine needles from which the snow had been pawed and scraped. The piglets may not have known we were there and gave no indication that we had alarmed them. The sow, on the other hand, obviously did not want us around and would have run us off had she been free to do so. She did make a series of short, noisy charges in our direction, but never moved more than 10 feet or so from her infants. Recognizing this safety margin, we edged to within 50 feet of the family to watch. The unsteady piglets pushed at leaves with their snouts, bumped each other playfully and nursed when the sow was not occupied rushing and swearing at us.

"She would fight for those little fellows until you killed her," Will said approvingly. "There isn't anything but death will make her leave them. I was thinking about some hippies I saw over in Bisbee. They had hair down to their tails and were all dressed up in leather suits and beads and cowboy hats. They had a little tiny girl, I guess it was theirs, with them. It was a cold day, but that little thing was barefooted and just wearing a pair of bib overalls. Her hair was all snarled. I'd like to bring her mama over and show her this javelina and let her think about it."

"Come on, Will."

"I'm serious. The first rule for everything, people or pigs, is take care of the little ones, no matter what, no matter what it costs. We've got a lot of problems because too many people have forgotten why that mama pig acts like she does."

Interested as we were, we decided that too much observation might create harmful stress for the pig family. We forded the stream, made a quiet, cautious circle around the javelinas and continued climbing in the canyon. When we returned in late afternoon they had moved farther back into the brush. The sow had cleared another nursery space in the snow and all seemed in order.

Up and down, we walked about 10 miles in Copper Glance canyon and by the time we got back to the truck we felt all of them. "I'm dragging," Will said. "I've spent too much time being a dozer jockey lately. But"—he uncorked one of his back-smashers—"what we've been doing is about as good a way as two old fellows could find to get tired."

The following day, 15-year-old Jimmy Sparks, Will's son, and I went off on an overnight trip, intending to traverse the main summit of the Huachucas and see what was abroad there. As far as we could tell, nothing was moving in the snow squalls and deep drifts at 8,300 feet. The last significant track we found was a good one, indisputably made by a mountain lion. The animal had hurriedly crossed the ridge through a col. We sensed that by the time the lion reached the top he had decided that whatever he originally had in mind was a mistake. We soon came to the same conclusion. At 9,000 feet the snow was four feet deep but still so soft and dry that even with snowshoes we sank into it up to our hips. After wallowing for a while we, like the lion, gave up and turned down to camp at the head of Sawmill Canyon. We packed out the next day without reaching the summit, traveling, in ecological terms, from Canada to Arizona in the course of a 4,000-foot descent. When we started, Jimmy had said he wanted to be an Arctic explorer when he grew up; when we finished he said he thought exploring jungles might be better.

On the last day I went up Montezuma Canyon to the old cabin where we had lived during the coati study. I was alone, as I preferred to be because of the ghosts and emotions the place arouses. At an old stock tank where we had swum and drawn water I found chulo tracks and then a chulo, an old male grumpily turning over mine tailings looking for lizards or whatever. He could have been a lonely survivor of the tribe we knew, one of the cubs of our day, or a pioneer, the forerunner of a new tribe coming to reoccupy the canyon. Either way it was a satisfying finale.

Going Under

Bath and Highland counties in Virginia and Pocahontas, Pendleton and Grant in West Virginia run together, a wild mountainous section that for promotional purposes has been called the Switzerland of America. Also it is, like Potter County, Pennsylvania, sometimes called God's Country. There are a number of other places for which proud residents make the same claim. Granting the First Premise the title is as valid one place as another. It is a becoming expression of environmental gratitude and satisfaction, not a mean, chauvinistic brag.

The tract has been settled for almost 200 years, but there remain large unoccupied sections, and the overall character of the area is distinctly rural. It is a place of small towns and villages (Petersburg, West Virginia, with 3,000 people is the largest community in the five counties), small farms (beef-grazing is the principal business) and small roads (west of Interstate 81 there is not another superhighway for 100 miles). This pattern of light, irregular settlement is dictated by the nature of the land. There are long ridges studded with peaks and knobs, cut by river canyons and locked valleys. There is heavy forest—Appalachian jungle growth of hardwood, hemlock, pine, rhododendron, laurel, greenbrier, grape and ivy. Early settlers simply lived around these obstacles. Technology for flattening them now exists but the rough country remains, often it seems because people are in the habit of living with it.

Purists might object to these highlands being called a wilderness, but then wilderness is a relative term. There is in nature no wilderness as there is wetness, dryness, coldness, a delta, a desert, an ice pack.

Wilderness can only be defined in terms of man—that is, a place which has only lightly and not too obviously been used by him. How much human activity denies a place wilderness status is a matter of taste. There are some preserves set aside by law and patrolled by rangers that are much more wards of man than the privately owned highlands.

As a small example, there is a ravine 30 feet deep and 20 feet wide that has been cut through limestone for a quarter of a mile by a big spring flowing to a tributary of the Potomac River. The ravine is hidden by enormous hemlocks. In a few places where light filters through, there are patches of watercress and cardinal flower. The mini-canyon is plastered with mosses and liverworts. On a face of rock near where the water flows out of the mountain there is a beautiful bed of walking ferns. This is a curious plant. It has slender fronds, much the shape and size of a bullfrog's tongue, which grow in graceful arches. Where the tips curve to touch the ground they take root and send up new plants. The walking fern is one of the rarest ferns in the Eastern mountains. It is finicky about growing, needing limestone, shelter and stable conditions of light, humidity and ground moisture. These ferns are better indicators than any legislative act or rustic sign that this ravine is wilderness.

The ravine divides two pastures of a farm owned by a family that settled there in 1797. The ravine was never timbered because it was not worth the trouble of cutting and dragging the hemlock out over the sharp water-pitted limestone. In the thicket there is little in the way of forage to attract cattle. Also the limestone has eroded into a series of knife-edged ledges, as full of holes as Swiss cheese. These act as a natural cattle guard. The ravine survived for a long time because it was too much bother to remove or reduce it. Now it is kept because it is of special value to the owners and a few others. The family picks watercress from the pools. Every now and then somebody walks up the ravine to admire the rock sculpture and the walking ferns or just because it is a nice place. An elderly member of the family says that on a hot summer day the ravine is more comfortable than any air-conditioned bank building he has ever been in.

In the highlands there are a good many places like the Walking Fern Ravine, which, if not wilderness, are treated as such. It is good country for rock climbers, white-water canoeists and cavers. There are some traditional spots where climbing cliffs overlook fast white-water streams, and here on a Saturday morning 30 or 40 pickups, campers, vans, 4WD vehicles and an occasional Triumph may gather from as many as 10 states. Enthusiasts of this sort may leave work on Friday afternoon and drive through the night to the highlands for a weekend of climbing or paddling. The highlands are a great place to learn these sports, but once they are mastered the region does not provide ultimate tests. For spelunkers, however, the situation is different. There is fine caving country elsewhere, but the best and most varied caving is in the limestone mountains of the two Virginias. The highlands are honeycombed with caverns grading upward in difficulty to places like Schoolhouse, Grapevine and Butler, which because of their beauty and risks are regarded as classic American holes in the ground.

There once was a cabin on the Bullpasture River in the highlands that was used as a kind of spelunking headquarters. Some weekends 50 people would use the cabin as a base. I remember one morning waking to find a Yale student hunched over a notebook, apparently having written through the night. "I cannot sleep when I am near caves," he said. "I am making entries in my journal."

Here are some made not long ago in mine.

Pancake's Cave

Caves are often named after the families on whose property an entrance is located. Pancake is a common name in the highlands. However, this particular cave is not located on Pancake land and the name given here is fictitious. (This deception and others used later are to make it more difficult to find the caves. Anyone looking for a cave or a bird or a walking fern will tend to respect it more if he must exercise a little ingenuity to find it, rather than having it located like a Holiday Inn in a guidebook.)

Friends took me to Pancake's Cave the first winter I caved. Since then I have returned every few years. The cave is medium-sized (about a mile and a half of passages) and of medium difficulty, providing some interesting crawls and scenery. Any reasonably active person can get through the main passages, but there are optional routes where rope, some skill and self-control are necessary.

The small entranceway is covered by a locked iron plate, the posts of which are set in the solid rock. The owner, J. W. Pancake, covered the entrance many years ago, not to be difficult but as an informal conservation measure. To explore the cave it is necessary to stop by the Pancake farmhouse and get the key, which J. W. sometimes withholds from those who strike him as being insufficiently experienced or responsible. It is a good system since it saves the cave from casual vandals and leg-breakers, and also gives people an excuse for talking with J. W., which is good for anybody. J. W. has a dog that waves good-by to visitors, bees that deposit their comb and honey directly in glass jars, three of the most responsive saddle horses you will see outside the Spanish Riding School and a barnful of clever inventions. J. W. is getting on in years but remains cheerful and entertaining and leaves one feeling better about our species.

Even the foyer, if you want to call it that, of Pancake's Cave has always seemed a pleasant place. It is an open pit, which long ago was probably part of the cave proper. The pit is overhung with the branches of big oaks and maples, which grow on the rim. Water seeps down the rocky sides and forms tiny pools where it is not uncommon to surprise deer drinking. The sides of the pit slope inward. At the very bottom, disappearing under the rock wall, is the small hole through which the cave is entered. The whole arrangement is like an enlarged model of one of the funnels that ant lions dig in the sand to trap insects. Because cavers, too, can find themselves trapped they are often asked, "Why do you do it?" The cool answer is, "Because it's there," which is not a totally false response but one that has been given so often that nothing more sensible seems possible. Caving is as explicable as, say, gambling, sightseeing or making friends. It is a less common activity than any of these, but like so many entertainments it is an attempt to find stimulation and pleasure by satisfying certain common urges.

Abstract curiosity, for example, is one of the definitive characteristics of mankind, a spur for all sorts of activities from philosophy to space travel. Caving may be only a trivial entry well down the list of things for which curiosity is responsible, but it is one. Take my last trip into Pancake's with a group of six. I had been down the entrance tunnel a dozen times. Terry had made the trip three times. The other five had not been there before but had heard about the place from us and had been in other caves. The point is that our curiosity was not virgin. Nevertheless, sitting in the pit, readying our carbide lights, adjusting hard hats, the small hole slanting downward into the darkness was physically and psychologically the dominant feature of the landscape; it probably would be for 999 out of 1,000 people.

Like a river, horizon or mountaintop, a hole in the ground stimulates curiosity. Primitive or civilized, young or old, brave or timid, it is almost impossible for a human to discover a hole and not wonder where it leads, what is down there. Having found the hole, there are many reasons why a man may decide to let it be—it is too small, it may contain poisoned air, snakes or trolls, the man had no light or time. However, if he passes it by, he is going to hurt a little because he could not or would not satisfy his curiosity. This is the elemental reason for caving—to experience the joy that comes from giving in to a deep urge. There almost never is a practical reason for going into a cave. The odds are very long against finding love, wealth or a good meal at the bottom of the hole. Caving is pure hedonism.

The entrance tunnel to Pancake's Cave descends sharply for half a dozen body lengths and then opens up into a wide, level passage that gently slopes into the mountain for a quarter of a mile. It makes a very nice entrance hall, a kind of decompression chamber in reverse. Without worrying about climbing, balancing or falling through a crevice, you can saunter along for a few minutes getting adjusted to underground peculiarities, the most obvious of which are darkness and silence. There is no light except that cast by the carbide flame of the head lamp, and except for an occasional drip of water and the noise one makes moving along, no sound. This is soothing because of the sharp reduction in number and variety of stimuli customarily bombarding our senses above ground. And because the stimuli are drastically cut back, the senses seem to deal more efficiently with them. You are hypersensitive to the shape of a stone under your foot, knowing immediately when and how it moves or gives and being very conscious of what this may mean. Uncommon awareness is one of the delights of caving.

The entrance passage ends abruptly, pinching down to a slit that disappears under the base of a tilted slab of limestone. This slit leads into a 100-foot crawlway. Crawling is the essential mode of travel in caves.

In some places one must crawl on all fours like an infant. But there are other passages too small for this, where one must flatten out on the belly or back and lever forward on one's elbows like an awkward snake. When they first go spelunking, four out of five people experience a certain amount of fright and panic, which for lack of a better word might be called claustrophobia. Most often it strikes in crawlways. Pressed against rock on all sides, inching ahead into absolute darkness under God only knows how many tons of mountain, it is almost impossible not to wonder if whoever built the cave was a careful workman and had the services of a good structural engineer.

Cave fright takes various outward forms: babbling, grim silence, sweats (both hot and cold), cursing, hysteria and involuntary rigidity. I remember one woman setting off on her first weekend of caving. Before the descent, she made some of us uneasy by talking far too much about how she knew she was going to love caving. About 50 feet into the first crawlway she began to whimper. At the next tight spot she froze as solid as an airline dinner. It was pointed out that bigger bodies than hers had gone ahead and that there were others behind who could not possibly advance unless she moved. None of this good counsel was effective. Therefore, the man directly behind the rigid lady took her legs and began to wriggle backward, trying to drag Ms. Muffet out of her tuffet. Her rescuer tugged so hard that her buckles and buttons popped and her jeans ended up around her knees. She instantly regained the use of her limbs and in fact began to scoot backward so smartly that the spelunkers behind her were almost crushed. Coming to a place where she could rise, she did so, drew up her pants and asked to be escorted to the entrance. She was. The moral is that cultural concerns—in this case, modesty—can affect behavior as powerfully as innate fears, such as claustrophobia.

The first crawlway in Pancake's Cave is a snug, squeeze-and-lever operation. It ends at the base of a 10-foot dry well, which must be climbed in order to get to an upper level where the passage continues. Chimneys are vertical crawlways, but this one comes at a good place, allowing one to stretch one's legs while climbing. After a bit the upper passage degenerates into a formation that spelunkers have named the Snakehole, although it is more a narrow crack between two smooth faces of limestone. At the top the crack is two feet wide, but it tapers downward so that at the bottom there is only six inches between the walls of rock. The only way to get through is to raise oneself, push-up fashion, to the top of the fissure. There, with back and butt pressed against the roof, you must suspend yourself, face, feet and hands dangling uselessly in the crevice, and press the rest of your body against the walls. Then you must squirm forward alternately shifting shoulders and hips, all the while staring at the narrows below, thinking if you slip you will remain forever wedged between the two walls like a rusty axhead in a chunk of oak. Well into the Snakehole another comforting fact becomes apparent: it is almost impossible to wriggle backward.

Long ago, some hero came to the entrance of the Snakehole for the first time and thought it might lead somewhere—it does; two-thirds of Pancake's Cave lies beyond—and ventured ahead. This act took enormous curiosity and may have provided about as much pleasure as the average man can stand.

Nobody is small enough to get through the Snakehole easily and for many it is a hard, scary trip. Naturally, those who have made it take delight in describing the place to those who have not yet been there. Veterans enjoy letting rookies go first while they sit around listening to their agony, giving advice and encouragement. In our party Beth and Sam were the biggest rookies, both six-footers or better. However, both are young and strong. There was some disbelieving palaver when the entrance to the Snakehole was pointed out, but they figuratively spat on their hands and did what had to be done. Mary Jane is not very big but she has spread out and softened since taking a newspaper job in Kansas. She came out of the Snakehole headfirst, red in the face, rock-burned but feisty.

"How did you like it?"

"First I was furious at you for getting me in there, sitting around waiting for me to squeak. Then I thought I couldn't make it. I wanted to squeak. My head knew I could make it if I got hold of myself. I did and now, if you must know, I feel quite good about myself."

Almost everyone has enough body (or, more accurately, not too much) to make the Snakehole. But the problem is not having a great body but coping with the one you have. It is a matter of grabbing yourself by the ears and saying, "Baby, behave. It is spooky but that is no excuse for choking. *I am in charge here and you are going to do it my way.*"

Self-possession is not only necessary but easy to experience while caving. The underground environment, the reduction of customary stimuli, the physical demands of climbing and crawling, the taste of panic make one acutely aware of the present moment. The problem is usually something tangible, such as how to get a hip around a four-inch projection of limestone. When you succeed, there is a sense that you are very alive, in good control of your functions. You are astonished at the strong and sharp satisfaction and realize how seldom you find the same goodness in more complex and conventional experiences.

A metaphysician-caver I know claims that caving is not a do-it-yourself substitute for psychotherapy; instead, a shrink is "a poor man's cave."

Windflower Cave

To get there from the nearest road, you have to walk through a pasture, then wade a shallow, swift tributary of the James River a few hundred yards above a spectacular gorge. On the far side you pass through a thicket of hemlock and rhododendron and then climb an open slope. In the spring this slope is carpeted with anemones and windflowers. About 500 yards above the river crossing you come to a shoulder of the mountain, on top of which you can sit, smell the flowers and look at and listen to the white water boiling below. And here is the squarish opening to the cave.

Windflower is a small cave consisting of a single main passage less than half a mile long. Generally this passage is 20 feet or so wide and the footing is good. There are a couple of squeezes where the walls pinch in but no true crawlways. In one of the narrow places the floor of the passage is covered with two or three feet of clear, cold water. On the way in, most people try to keep dry by straddling the water, hunching along above it with feet and hands braced against the walls. Invariably somebody slips into the water, and the party gives in to wetness and wades through the underground pool.

Walking through Windflower Cave takes about as much skill and involves about as much risk as walking on a mountain path. Nevertheless, a lot of good cavers enjoy visiting the cavern because it is such a beautiful place.

Cave scenery is very different, principally because it is so simple. There is no vegetation, no wind, no clouds, no shifting source of illumination, no life (except for a few blind insects and amphibians) to distract attention from the rock and water sculpture. But the results of water flowing over, under and through limestone, dissolving and redepositing it, are infinitely complex and surprising. In Windflower Cave limestone turrets, wedding cakes, toadstools and sponges grow from the floor, hang from the ceilings and sprout from the walls. There is a section of ceiling covered with soda straws—fragile, almost crystalline tubes of hollow limestone through which absolutely pure water drips slowly. One of the gallery rooms is 60 feet high, and its walls are adorned with graceful, fluted, filigreed ledges. As illuminated by carbide flame, each formation has a different color value and tone, from milk-white through gold and tawny to obsidian.

The impact of the strange color and form is heightened by the perspective. Rather than looking down, up or across at them, as is usually the case aboveground, cave formations are seen from the inside, which creates a three-dimensional effect. There is a surrealistic illusion of great space. In the main gallery room the fancy limestone work quickly soaks up the light from a carbide lamp. The limit of vision is 50 feet. Beyond the rim of flickering light, beyond the last shadow, there is total blackness. In fact, just beyond the light there may be a solid rock face, but the darkness suggests immense reaches of space.

A notable feature of Windflower Cave is the Bacon Rind Room. Here, on one wall, a hundred peculiar limestone slabs grow. Each is only an inch or two thick, vaguely circular, as much as 10 feet in diameter. They are clustered rank on rank and look like the gills of an enormous petrified fish. If tapped gently each gill gives off a slightly different, vibrant, organlike sound. The musically inclined can make eerie underground music on these gills. John made our music. "Wow," he said, listening to the echoes bouncing off the irregular walls, "if I could take this scene with me, I could absolutely blow the mind of every head in Isla Vista." In many caves there is what might be called aesthetic food. There is nothing extraordinary or eccentric about the satisfactions to be found in a place like Windflower Cave. Rather it would be a profoundly unnatural man who could not find enjoyment there.

Dead Horse Cave

Pancake's is subtle; Windflower is delicate. There is very little of either quality about Dead Horse Cave, a great brute of a hole with 10 miles of passages that branch off in a bewildering maze. There are circular walks and crawlways and deep wells and high chimneys that connect half a dozen different levels. There are several enormous bowling alley-sized rooms. These are impressive formations that are more massive than intricate. Within the cave, which is dry, there is an abundance of scree and rotten, crumbling ledges. Many passages and rooms are littered with great blocks and shards of rock that have fallen like pieces of giant tile from a decaying mansion. All in all it is a place that makes many, including myself, vaguely uneasy. It seems as if it is the kind of hole trolls might inhabit.

The day we decided to inspect Dead Horse we were running low on carbide. We had only enough left for five of us to stay underground for three hours. I was not annoyed about this shortage, which in fact fitted in nicely with the doubts I was having about Dead Horse. I alone had been in the cave, but only twice, and the last time 15 years previously. I was certain that I could not sort out the complex passages. What with carbide and confidence being in short supply, I suggested we try to do only one thing: find a big room playfully named Grendel's Pantry by spelunkers. It lay, I thought, within an hour of the entrance and could be reached without too much trouble.

We floundered about for a while probing dead ends before emerging from a crawlway into a passage that seemed headed in the right direction. This passage was deep, narrow and multilevel. Every so often it was interrupted by a wall of rock, a gaping pit or chasm. It was necessary to climb over or go under these obstructions, ascending a chimney or dropping down through a well and continuing on at a different elevation. Some of these downers and uppers were difficult, requiring traverses of flaky ledges that overhung free drops of 30 feet.

Through no failure of skill or nerve, Beth had taken a short but painful fall in Pancake's Cave two days before. She had bruised her back and had stiffened up considerably. Soon we came to a broad ledge, where we sat down for a break. Beth asked if we would come past this place on our return. I said we would, that we did not have enough carbide or a good enough guide to look for alternative passages. "Then I guess I'll stay here and wait for you," Beth said. "I'm sore and I'm getting a little shaky."

Beth is a strong, venturesome woman, a national-class athlete and physically the most talented of all of us. Being this and a woman, she is extremely sensitive to any suggestion of frailty or timidity. Knowing Beth, we were impressed by her decision. Self-possession takes various forms. One is knowing what you can do if you take hold of yourself firmly. Another is knowing what you cannot do, the limit beyond which mind-over-matter games become dangerous fantasies.

A quarter of a mile past where she stopped we climbed toward the top of the canyon room and came out on a wide ledge where we could walk upright. We then began to encounter shafts angling up through the roof. I investigated these, having the recollection that the final entry into Grendel's Pantry was made through such a chimney. Most of them were obviously not right, extending only a few feet upward through the rock, but one that kept going beyond the range of a headlamp seemed promising.

Chimneying is a technique for ascending an open shaft or advancing horizontally between two faces of rock. It involves bridging the open space with your body by bracing your shoulders, back and hips against one wall, your feet against the other. Pressure keeps you suspended in place, and by shifting your body and the pressure you move ahead. It is not a difficult or strenuous maneuver. Rock walls are seldom smooth. There are projections and niches on and in which you can take a breather.

This particular chimney was big and easy. It started upward to the left and after about 15 feet bent sharply to the right around a knob before continuing for another 20 feet. The top seemed to be wedged shut with assorted slabs, but I went all the way up and braced in the blocked area to make certain that there were no passable crevices. While so engaged I felt a hint of movement, a certain uneasiness in the rock under my left foot. Swinging my light around, I saw I was braced on a roundish chunk jammed in between the wall of the chimney and, above, a really formidable 6'x6'x6' block of limestone, which, if it ever came down through the shaft, would make an effective tombstone. That the small rock was not solid was obvious. How important it was to the stability of the tombstone was not at all obvious, but when the possibilities became apparent, I felt an instant stab like an electric shock starting in my lower stomach and traveling upward rapidly. In response to this flash of panic I started down the chimney, very gingerly at first but swiftly once I had rounded the midway knob. I fell out of the bottom in a shower of scree and sweat.

"What's the verdict?" Sam asked.

"There is a rock up there like a pointy grapefruit which may be holding up another one that weighs approximately eight tons."

"Leave us not linger," advised Sam, and we did not.

We went back and collected Beth, then worked our way to the surface. The group seemed to be feeling very lively. Like kids dashing through a sprinkler trying to avoid the water but not really minding when they didn't, we had done a little flirting with fear.

Without being entirely perverse or contentious there is something that can be said for fear. Pure physical fright can be a tonic, like a very cold shower or 150-proof rum. A wild cave is a fine place to taste this. Darkness and claustrophobia provide a kind of substratum of fear underlying even routine caving. Then there are special circumstances—the prospect of getting lost, falling into an abyss, becoming the permanent cornerstone of a cave—to give extra jolts of stimulation.

Even if all of this can be momentarily accepted—that caves are scary and a little terror can be kicky—there remains the question of why bother to go into a hole or scale a mountain or run rapids to get this thrill. After all, ordinary, everyday life increasingly provides abundant opportunities for being scared to death. Perhaps it is that there is an unsatisfactory abstract quality to the worst contemporary alarms. Poisoned air, rich food, crooked politicians, high taxes, paranoid bosses, predatory junkies and highway traffic are far more threatening than the worst cave chimney. Each day we run the risk of being victims of something or somebody. But these do not cause stabs of pure fear as often as they do insomnia, ulcers, overeating, drinking and smoking. Worry and anxiety seem to stain and deaden the soul while genuine old-fashioned natural fear tends to whiten the teeth, sweeten the breath and make childbearing a pleasure. Faced with the prospect of being ultimately conked by a boulder, a shriek goes up, to be sure, but it is an affirmative one—"Hey, I'm alive, and very glad of it."

Journey into Spring

In north-central Pennsylvania there is a complex of connected rivers on which I have had a lot of good times during the past 20 years: the Sinnemahoning and Kettle flowing out of the Black Forest; Pine Creek rushing through the magnificent Grand Canyon of the East; the Moshannon and Loyalsock, pre-eminent white-water runs. These, as well as hundreds of other streams, creeks and springs are tributaries of the West Branch of the Susquehanna River.

There is a something about a river that invites you to travel it from beginning to end. Considering what high regard I have for the West Branch, it may seem curious that I had never made such a trip on this particular stream. However, there was no sense of deprivation or frustration, because I felt certain that someday the time and circumstances would be ripe. And sure enough, time and circumstances coincided last spring. Lyn, my eldest daughter and a college junior in Arizona, called one April evening to say that she had a month off between the end of the spring semester and the beginning of summer school and that she was coming back to Pennsylvania. "But I'll get bored if I just sit around playing with toys and people I used to know," she said. "I want an adventure. I was thinking of . . . ."

"What I have been thinking about—again," I said, "is paddling the whole West Branch. You want to do that?"

"That's weird," Lyn said. "I was going to say before you interrupted that I've really been thinking about a canoe trip. It's been 10 years since I was on a real one."

"When can you start?"

"My last exam is on May 12. If I fly, I'll be home the next morning, but if I ride the bus it will be later."

"Make a reservation. I'll send a check for a plane ticket in the morning."

Like old hemp ropes, rivers tend to fray at their upper ends, dividing into a number of more or less equal strands, any one of which can logically be called the source. This is the case with the West Branch, the original strands of which rise in a 100-mile arc of highlands that stretch from roughly the Altoona-Johnstown area to the New York border. Through the offices of the U.S. Geological Survey, one of these strands has been designated as the West Branch. It begins in Cambria County on the outskirts of the village of Carrolltown from a number of seeps in a swampy bowl overlooked by several junk-food drive-ins and a public riding stable. It is not a picturesque scene, but humble beginnings should not be held against anyone, or any river.

From there the West Branch runs north and east for 240 miles—through mountains for most of the way—until at Northumberland it joins the Susquehanna, which rises at Cooperstown, New York, out of Otsego Lake and flows southward through less rugged terrain. The Susquehanna rolls down through the Harrisburg-York-Baltimore megalopolis until it enters—in fact, creates—Chesapeake Bay.

All in all, the Susquehanna drains 27,500 square miles. It is the largest river flowing into the Atlantic Ocean from the United States. Each day an average of 23 billion gallons pours out of the mouth of the river. However, this enormous total is more or less a drop in the bucket compared to what the river can do when it is really trying. In the flood of '72, hydrologists calculated that at its peak the river was discharging nearly 700 billion gallons of water a day into the Chesapeake.

The West Branch accounts for 6,990 square miles of the Susquehanna drainage. When it floods, it can have a quick and vicious temper. I have seen August thunderstorms send it raging over its upstream banks. However, it has been less destructive of human life and property than some other components of the Susquehanna simply because there are fewer people along its course. This holds true especially in the up-country, which is sparsely populated; in many areas there are fewer people than there were a century or more ago, reflecting the gradual depletion of the basin's two principal resources, timber and coal. All through the West Branch country there are corduroy logging roads now rotting and being converted into compost; old mine shafts being filled and broken by slides; breached millraces that are now being used only by trout; collapsed bridges traversable only by raccoons; villages that were once prosperous towns. The West Branch basin of the Susquehanna may not be the best place in the world to be if you have to earn your living there, but it is a great place for wilderness and all the things that go with it.

In terms of topography and history, the obvious place for beginning a descent of the West Branch is the village of Cherry Tree, about 30 miles from the springs that are its official source. Cherry Tree was something of a boomtown during mid-19th-century logging days, being a convenient place to commence driving logs downstream and for rafters and jacks to carouse when they were not working in the woods or on the river. It is still a pretty, shaded village, but the river is merely a 30-foot-wide stream, badly silted and muddied. Quite obviously, Cherry Tree now has not much regard for the West Branch except to use it as an informal, and probably illegal, dump. And, except in times of flood, the stream here does not hold enough water to float a canoe. By ten miles or so below Cherry Tree, at McGee's Mills, the West Branch has received several infusions of water from tributary creeks, and it has leaped over rocks and gravel bars, rolled on and aerated itself many times. Thus the river has been purified.

McGee's Mills is another former logging village, but one which has shrunk to a hamlet consisting of a few old houses. Among the remaining 19th-century structures is a covered bridge that no longer bears traffic but is retained for its charm and antiquity. A lane leads down to the bridge, and the area around it is an informal park shaded by big pines — a nice picnic spot and fishing site. It is also a very good jumping-off place. The banks slope gently down to the river and it is not necessary to dodge traffic while unloading gear. However, the bridge is not far off the beaten track. As Lyn and I unloaded our gear, down the lane rattled a modishly painted van out of which emerged a fashionably bearded, sandaled, bandannaed young man who was carrying a three-foot pine plank. He immediately said, "Wow, far out—a canoe. I'm into martial arts in Altoona, but I had a grungy week. I told the old lady I had to get out in the country and get my head together. I ripped off this board and I get anybody I see doing anything cool to sign it. You have got to sign."

"What are you going to do when it's filled with names?" I asked.

"Split it with a single karate chop?"

"Hey, man, be serious. This is going to be one of the best things I own. I'm going to hang it up, and when I get down I'll look at it and think about today and the people and that will bring me up."

From McGee's Mills to the mouth 200 miles away, the West Branch, except for a few roiled and poisonous stretches, is crystal clear. It is so clear because it flows over rocks and because there is not much loam or farming land or living room on the ridges it passes. Also, in certain places the clarity comes from impurity, from infusions of acid mine waste, which kills organisms that make water murky.

About 30 miles below McGee's Mills the West Branch is a big trout stream. It is very shallow and crooked, writhing against cliffs, wriggling through gaps, around boulders, gravel bars and ledges. Riffles are almost constant, and there are some fast chutes and short drops but no formidable rapids. It is entertaining but not dangerous white water. All of the problems met in heavy water are there, and all of the downriver techniques can be practiced. However, the penalties for error are not very harsh. In big white water, if you select the wrong stroke, miss an opening in the rocks or try to force maneuvers in the current, you may be swamped, even drowned. The shallow West Branch reminds you of your mistakes, but gently—you get a cautionary bang from a rock on the keel or at worst end up stuck on a ledge and have to get out and pull your canoe free.

Lyn said at the start, "I haven't paddled on white water for a long time. Maybe I'm not so good any more. If you yell at me, I'll yell back."

"You'll be all right—you had such superior early instruction," I said modestly. "Anyway, I never did yell that much—just coached forcefully."

In white water the bow paddler should be the lighter one so that the bow rides higher than the stern. This has led to the common but fallacious assumption that the bow position is somehow inferior to the stern. The truth is, in white water the bow paddler, having the best vantage point, is the tactical commander, picking the best route through the turbulence. The bow is responsible for getting the first half of the canoe through an opening into a chute or over a rapids. The stern must then follow with the second section. It is not as simple as it may sound, which is one reason in heavy white-water courses you often see abandoned canoes, bent at right angles around rocks. Improvising always, the bow throws strokes back to his or her partner, who tries to catch them, follow and adapt. It is very free-form maneuvering—in some ways like dancing—with almost no repetition, because the beat and temper of the river are never exactly the same in any two places. White water can and should be paddle-danced without a lot of chatter, strokes following observation and feel rather than commands or comment.

As it happened, Lyn had forgotten very little and, more important, had indisputably retained a certain aggressiveness of personality, which makes for a good bow paddler. She began taking charge, reading the water, decisively picking holes—usually the right ones—taking what the current and river topography gave us. After a day or so of adjusting in the shallow water upstream, we were working nicely by the time we hit bigger rapids in the downstream canyons. And there was no yelling.

For me, memories of particular places are usually as much aural as visual. One summer I walked the length of the Appalachian Trail, and one of my most vivid recollections is the song of white-throated sparrows, heard everywhere as I moved north with the spring from Georgia to Maine. Northern lakes are loon cries. The tundra is the whine of mosquitoes; Los Angeles is the roar of traffic, Las Vegas the clank of slots.

On any white water there is a basic sound of current slapping and swooshing against rock, but the special sound of the West Branch, the one that I suspect both of us will always associate with that trip, was of chirping chipmunks. I have no zoological data to support the claim but it may be there are more chipmunks along that river than anywhere else in the world. They are there because there are tons of acorn and other mast and an infinite number of superb den sites in the pockets of humus between the rocks and roots. In part because it is such a good place for chipmunks, the narrow West Branch vales are hunted by hawks, owls, foxes and other predators.

One morning we tried to make a crude estimate of the chipmunk population, counting all the little ground squirrels we definitely saw or heard while floating along a mile of riverbank. A chipmunk every 20 feet seemed to be the situation, but these were just the ones lined up on the shore. Behind them, extending back to the cliff, were additional ranks. And presumably they were as numerous on the other side of the river as on ours. All in all, 5,000 chipmunks a mile, a million or so in the 200 miles of river, did not seem exaggerative. Whatever the true number, we were seldom out of sound of their chirping, which is very nice music for traveling.

One great convenience of river travel is that there is seldom any need to plan ahead, to push on to find a desirable stopping place. In the entire 200 miles of the West Branch we paddled there were probably no more than 30 that were not suitable for camping. Late on the first afternoon we came around a bend, and ahead, on a mini-delta created by the outflow of a tiny streamlet, we saw a doe and two fawns drinking. Above the gravel bar was a flat, clear bench shaded by hemlocks and sprinkled with blooming trillium. The deer ran off but we stayed there for the night.

"Exit at the second fawn. Turn left at the trillium patch," Lyn commented. "What a wonderful way to end the day."

There are five dams on the West Branch. The lower four are minor obstacles around which a canoe can be carried without much inconvenience. The fifth — the first met upstream, at Curwensville — is a brutish structure and it requires a brutish portage: a 150-foot, 45-degree ascent of the breast, which is made of loose riprap liberally mulched with broken beer bottles, then a half-mile descent to return to the river. This takes an hour of sweating and cursing.

The Curwensville dam is about 130 feet high and half a mile long; it creates a seven-mile lake. The dam is the work of the Army Corps of Engineers. Like every canoeist I know, I am not fond of portages of any sort, and man-made ones always seem to be the worst. In strict recreational terms, therefore, I am not a fan of the Army Engineers, our leading portage makers. Also, like most canoeists and environmentalists in general, I am suspicious that if given their head the Engineers, like so many crazed beavers, might stop up every bit of flowing water we have.

The year before, I had called on the Engineers' office in charge of good and bad works in the Susquehanna basin. I had heard a rumor that a dam was being planned on the West Branch near a place called Keating. If it were to be built it would create an impoundment that would drown a section of canyon I particularly admire. I think it is of special natural significance. Indeed, said the Engineer informant, some studies were in progress. Eight or nine Susquehanna dam sites were being investigated, including the one at Keating. However, not to worry. As this Engineer explained it, Engineers spend a lot of time and money investigating the feasibility of dams. Potential hydroelectric, flood-control and recreational uses are considered and assigned numerical values. These must add up to a plus factor—*i.e.*, the dam must have some utility, at least on paper, before blueprints are drawn or concrete poured. In the case of the proposed Keating site, the preliminary utility figure had come up negative: there were more reasons not to build it than to build it.

I still believe the corps deserves close and constant watching, even harassment, but I am not as adamant as I once was (though many still are) that undammed rivers are always better than dammed ones, and that this is a morally superior position. It seems to me that such rigid value judgments are environmentally unnecessary, and unbecoming.

Curwensville's essential purpose is supposed to be flood control; secondarily it provides recreation. Lyn and I did not especially enjoy the seven-mile lake, since slack water is always a drag after you have been using and enjoying fast water. In the lake we had to dodge a number of powerboaters, fishermen and water skiers. Finally there was the ornery portage. Despite all this, Curwensville is an attractive lake, and if it had been made by, say, glacial action rather than by the Engineers, it would be widely admired both for its beauty and for the wildlife habitat around it, not to mention the fish within it. All the powerboaters seemed to be enjoying themselves in ways they could not have in a shallow, free-flowing river. Their sport was not ours, but they did not appear to be noticeably less sensible or sensitive than Lyn and I.

Nine more dams on the Susquehanna would be awful and excessive.

Even one more at a place like Keating would be a major loss for me, and I think a major one for the country. However, Curwensville Lake does not absolutely corrupt the whole West Branch for me and it obviously enhances its value for others whose pursuit-of-happiness rights are just as valid as mine. The boaters, fishermen and skiers make up a potent pro-clean-river constituency. It is neither seemly nor wise for environmental purists to be snobbish about them because they have different esthetic and recreational values.

Clearfield is the principal upstream commercial and trading center on the West Branch. The river runs directly through the town—Clearfield is the only place I know of where you can pull up a canoe in an A & P parking lot—and on the whole has been treated well by it. There is about as little debris as it is possible for 10,000 persons to create. The water remains clear, suitable for swimming, very suitable for fishing.

Downstream, things are much worse, for a few miles worse than anywhere else on the West Branch. First there is the town sewage plant. Wastes are admittedly treated but then are spewed into the river in a great milky, odoriferous gout. Next to the sewage plant is the mouth of Clearfield Creek, a tributary that is almost as big as the West Branch itself. The creek is clear, but suspiciously so, being an odd yellow color like liquid topaz. The rocks in it and the banks along it are stained a startling orange. The color comes from iron sulfate, which is dissolved in abandoned coal mines and pits, then leaches down into the streams.

At this point there is a strange and very visible mix. Lyn, who had never seen such a concoction, called it a Triple-S cocktail. The yellowish water of Clearfield Creek runs in a distinct channel down the south side of the river. In the middle is the flow of whitish, opaque sewage. Most of the clean upstream water hugs the north bank, but it is shortly overwhelmed as Sulphur, Sewage and Stream are quickly blended, with immediate and toxic effect.

Upstream in the clear water we had seen a lot of fry, minnows and bigger fish. In the shallows along the shore there were great balls of polliwogs, and further inland an almost continuous line of frogs. Also present were two biological indicators that invariably attest to the liveliness of water. Sandpipers skipped about on the rocks, even in the middle of rapids, foraging for aquatic invertebrates. Kingfishers perched and hovered in the air, and then plunged into the pools for fish.

Very suddenly, at the point where the Triple-S cocktail was blended, all of these things disappeared—fish, amphibians and birds. The last living thing we saw as we approached Triple-S was a small sucker that had been washed into the mess and was swimming feebly on his side at the surface.

As well as any place in the country, the confluence of Clearfield Creek and West Branch makes an irrefutable case for environmentalism. Poisoning a river—the major artery of an enormous biological community—is a terrible, destructive and dangerous thing to do. Yet, paradoxically, the situation below the point of poisoning also demonstrates how potent and successful a force environmentalism has become.

Fifteen years ago, when I first began visiting the West Branch, there were about 100 miles of virtually lifeless water below Clearfield. Numerous communities and individuals were pouring raw sewage into it and most of the tributaries were scarlet and scalding with acid mine waste. Now, six or seven miles below the poison point, we saw our first downstream fish—again a sucker—and, coincidentally, at the same spot we also saw the first two sandpipers. There are, even now, no big fish for the next 75 miles or so, but the foundations for fish and fishing are being reestablished. Big beds of mixed aquatic vegetation have begun to grow in the river. Invertebrates are reproducing and feeding in them. There are schools of minnows at the mouths of pure-water tributaries, more raccoon tracks in the flats, more wood ducks, mallards and kingfishers nesting along the shores. Untreated sewage is now a minor, if essentially criminal, problem. Acid still enters at Clearfield and some other creeks, but many of the tributaries have been purified, and overall the river acidity is less than half of what it was.

Our third day we met another pair of canoeists—Gus, a young engineer at the Piper aircraft factory in the downstream city of Lock Haven, and his father. Both are longtime residents of Lock Haven, and both have spent a lot of time on the river. We got to talking about how it has changed for the better within our memory. "We have some bass now at Lock Haven, some trout a few miles below," said Gus, "but the biggest thing I notice is the swimming. I swam in this river as a kid, and I guess I'm lucky to be alive, considering what was in it. You could only stay in

10 or 15 minutes because the acid burned your eyes so bad. Now you can snorkel or swim for hours without irritation. It is better water than you get in a lot of motel pools.''

The river itself has contributed mightily to its own cleansing, but it is that complex of concerns, issues, laws and forced and voluntary actions we call environmentalism that has given the river a chance to rejuvenate. Twenty years ago the poisoning of a river did not matter to many and was accepted as an inevitable consequence of our style of life. Now it is regarded as very bad business, as both a malicious antisocial act and an unnecessary one.

The restoration of the West Branch is by no means complete. Nor, now that it has begun, is the resurrection assured. If we grow complacent or miserly about environmental works, the cycle can and will be reversed. However, this is less likely than it was even 20 years ago. We have accepted the fact that, as a kingfisher is a barometer of biological conditions in a river, the condition of the river itself is an indicator of the quality of human life.

About 10 miles below the Clearfield sewage works, the West Branch commences a mountain passage through the Allegheny Front. For most of the next 70 miles it is squeezed between ridges that rise 500 feet or so above the water. The riverbanks are covered with dense stands of hemlock and rhododendron and above them is a largely unbroken mixed forest of pine, oak, maple, cherry and birch. In this stretch there are only two tiny villages, Keating and Karthaus, and only a dozen or so other permanent dwellings on the river. There are only three bridges in the canyon area and no dams. There is no other wilderness so extensive and isolated in the Susquehanna basin, and not many anywhere that rival it.

I have been coming to parts of this canyon for a long time, but these few days in May were the best ever.

The best water. There is still too much sulphur in it and not enough life, but it no longer smells like acid, and, as noted, everywhere there are signs of returning life. Otherwise the water is superb, classic West Branch clear. Deep enough so that there is no dragging or hanging up on ledges. Fast enough, with big rapids for entertainment.

The best weather. An unbroken series of dry, sunlit days and clear, cool moonlit nights.

The best company. A daughter on the safe side of the generation gap and adolescent trauma who has become a companionable young woman.

The best entertainment. Sport in the white water, swimming in the deep pools, sunning on flat rocks, cribbage by the fire, and everywhere, always, an immensely varied display of geological, botanical and zoological wonders.

Beginning with the mode of travel itself—improvising as the current demands—there is something wonderfully extemporaneous about canoeing on a wild river. Each day you know you are going to come across interesting phenomena, but you have no idea what they may be or how they will be met. There is an urge to get on with it, to see what the river and canyon have to offer.

A red fox vixen at the mouth of a den between rhododendron roots, staring intently as we drift under her bank without moving our paddles, or even our eyes. . . . A clump of azalea bearing perhaps the most brilliant flowers that ever bloomed, so bright that at a distance, illuminated by rays of the early sun burning through the river mist, the blooming bush is mistaken for a flame. . . . A black bear track along an old corduroy road, which the bear was systematically demolishing as he hunted for grubs, mice and chipmunks. . . . A bed of shale studded with fossils at the mouth of a big tributary stream. . . . Two young great horned owls, their heads still fuzzy with infant down, unsteady and clumsy on wing. They are accompanied by an adult on what may be their maiden flight. It probably is not a pleasant one for them. They are picked up by a mob of cursing crows. As we keep pace on the water, the big hunters are harassed and driven half a mile downriver from hemlock to hemlock.

There are a few acres of flattish land gouged out of the canyonside like a terrestrial cove. This land is marked on old maps as Gallows Harbor. It was cleared and occupied by someone. Homesteaders? Loggers? The military? Now all that is left of the settlement is a wild meadow, the crumbling foundations of several cabins and a dug spring, the rockwork of which is smooth with moss, dripping with columbine. We drink at the spring, fill our water bottles and wonder about who lived here, particularly about what the ominous name Gallows Harbor signifies.

We never think about future ghosts. We can speculate about who was here a century ago, but not about who will be here a century from now. Once I was planting walnut trees at home and thinking about who would pick up and shell the nuts. I couldn't see them any more than the men who dug this spring could imagine us drinking from it.

Every canoe tourist I know has firm notions about what constitutes the Perfect Camping Spot. The PCS begins with a flat beach or wide shelving rock to which you can draw directly alongside and unload without wading or slipping in the mud. The natural wharf is also good for sunning upon after you have gone swimming in the deep pool that lies in front of it. Above the beach is a level bench covered either with evergreen needles or a heavy mulch of dry leaves. It is spacious enough for laying out bags, hanging a drying line, pitching a tent and building a fireplace (for which suitable flat rocks are nearby). Within 50 feet there are a pure spring, birch bark for tinder, dead hemlock for kindling and quantities of fallen hardwood that can be broken over the knee for the fire. Finally, on this bench, situated so as to give a scenic, contemplative view of the river, are two or three trees that have been bent backward by floods and whose trunks therefore make comfortable backrests.

Not infrequently on canoe trips I have become involved in semiserious squabbles about when and where to stop for the day and set up camp. There is no disagreement about the nature of the Perfect Camping Spot, only about whether a spot in question is as near to perfect as you are going to find that afternoon. The problem is compounded in the West Branch canyon, where there is an embarrassment of PCSs. There are so many that it is difficult to choose, and they are so perfect that it seems criminal to pass any of them by, even in midmorning. The West Branch canyon would lend itself to a very satisfying project—camping at every PCS in it. A trip for this purpose might take the better part of a spring and summer.

We camped one afternoon in a PCS that had a big spring out of which ran a considerable drain that curled around the campsite through the hemlocks. Leading down from the bench to the run was a well-used game trail. Astutely analyzing the signs, I predicted that this PCS might be a good place for mingling with beasts. I was right.

Pennsylvania has more white-tailed deer than any other state except Texas, and the West Branch basin has more deer than any other area in the state. We saw lots of deer every day in and around the river, and we saw even more at night. This particular evening they were thick. Having brushed away several families of jumping mice and chipmunks, tying the food bags on limbs out of reach of raccoons, bears and whatnot, I retired to a little knoll above the spring drain. As soon as I doused the lantern the deer began to move in out of the brush where they had apparently been waiting.

Deer, especially nighttime deer, do not entirely deserve their reputation for grace and surefootedness. Three or four of them stumbling about in a small campsite, tripping over ranks of firewood and their own hooves, can make a considerable racket. And in their excitement they become very vocal. In a state of mind somewhere between alarm and puzzlement, deer make a peculiar sound. It is quite un-deerlike, and I have heard those familiar with it claim that it was coming from mountain lions, wolves or maybe sasquatches. To me it has always sounded something like a heavy smoker waking up with a cold and a hangover. It is a combination of snort, hiss and bark, which can be described as a very loud "snisk."

The herdlet at this camp was especially noisy, and one animal circled me, snisking so raucously that it seemed he might be working up to a tantrum. To get a little quiet and to avoid becoming a living trampoline, I picked up my sleeping bag and moved inland, lying down beside the tent in which Lyn was sleeping so as to have real protection from mosquitoes and imaginary shelter from bears. Shortly after I had resettled, something else joined the party. From the sound alone it seemed to be a set of asthmatic venetian blinds. I flicked on a light, directed it toward the grunts and rattles, eventually illuminating a stout porcupine making his way ponderously but surely toward a salty skillet. I called Lyn, and finally she poked her head out of the tent flap. At once she began to laugh at the porcupine, who deserved it. On being discovered, he turned his back, threw up his flat tail, spread his quills and began to rattle in what he must have thought was a menacing fashion. Every few minutes he would look over his shoulder to see if he had frightened us, and finding he had not, would resume the official scare stance.

"Look at him," said Lyn, giggling. "He looks as if he's thinking, 'Oh bother, now I have to be fierce again.'" Head on, porcupines often look like reform candidates for mayor of Pittsburgh.

By and by we herded the porcupine out of camp. The sniskers quieted down and we finally got some sleep. The next morning, when we were talking about the events of the evening, Lyn came down with another laughing seizure. "I wasn't going to tell you, but I have to. It's so funny—about why you had to keep yelling to tell me the porcupine was here."

"Why?"

"I'm not scared of the animals but I am scared of noises at night. If I get thinking about them I can't sleep, so I stuffed Handi-Wipes in my ears. I could just hear you, but I didn't want to take them out because I thought you had a bear."

Lock Haven is a town of some 11,000 stretched out along the West Branch. Swimming, fishing, floating on inner tubes and sitting along the banks in gazebos seem to be popular entertainments. Lock Haven has a state college, the Piper aircraft works and probably many other points of interest, but from the restricted view of itinerant canoeists the best thing in town is the Fallon Hotel.

The Fallon has a long and exotic history. Spain having sold Florida to the United States, Queen Maria Cristina had some ready cash. This came to the attention of two Irish brothers, John and Christopher Fallon, who were—how shall it be put?—international entrepreneurs. In the 1850s the Fallons convinced the queen that building a fancy hotel in Lock Haven would be a splendid investment. Maria Cristina, bless her, agreed to bankroll the Fallons, who put up a big baroque structure with brick walls, high ceilings and gaudy chandeliers. It was the talk of the river.

In recent times the Fallon has been modernized and toned down but it is still an impressive building. An annex has been added at riverside. Guests can step directly from their room onto a wharf-patio and sun themselves, or go swimming or go boating. This is also the best commercial canoe landing spot on the West Branch.

Paddlers are not a large percentage of the Fallon customers but they appear often enough (another pair had stopped by just the year before) so that the staff is not alarmed by them. No rude remarks are made to guests who drip water on the carpets. A bellman will assist with muddy duffels, and a desk clerk advises that since the security of parked canoes cannot be guaranteed, it is advisable to take paddles to the room. All in all, the Fallon is a great place to catch a hot shower, nourishment from the salad bar and some soft sheets, quite in keeping with the Perfect Camping Spot tradition of the West Branch.

Below Lock Haven, fed by such fast-water tributaries as the Sinnemahoning and the Pine, the West Branch changes from a mountain stream to a big valley river. The ridges recede and farmland becomes conspicuous. The first morning out of the Fallon, opposite Great Island, we saw cows for the first time on the trip. It was somehow appropriate to meet them there. Two centuries ago this big island served as a ford for herds of migrating wood buffalo. According to witnesses the sound of the big herd bulls bellowing challenges at their rivals was deafening around the island crossings.

Good camping places are less frequent on the lower shores, but fortunately the river has become of a size and temper to create a number of isolated islands, too vulnerable to flooding to be of much permanent use. One of these on which we stopped was half a mile long and had a classic river-island profile. The upstream end was still growing from accumulations of silt and debris. It was low, marshy and covered with wet thickets of mallow, sedge and river-birch saplings. As we drifted along its shore the land became drier, firmer, higher and older. The downstream end where the first land had been made was 30 feet above the water, with a sharp bluff at the very tip.

There are seven or eight level acres on top of this bluff, and they support an extraordinary variety of trees, some growing more than 100 feet tall, representing both highland and lowland forest types. In this small area we identified four species of birch—gray, yellow, black and river—three species each of hickory and maple, two oaks, walnut, black cherry, ash, tulip, poplar, basswood, sycamore, locust, cottonwood, elm, sassafras, white pine and hemlock. Such diversity and proximity is unusual, since many of these species are not generally compatible. Big oaks, maples and tulip poplars, for example, will shade out and eventually kill smaller trees, evergreens and sun lovers such as locust and sassafras. The fertility of the moist but well-drained island soil contributes to the special circumstances. And the island is narrow—50 yards or so across—which permits light to enter from both sides rather than only from above, as would be the case on the mainland. The trees must have taken seed at about the same time on the island—from the looks of the bigger ones, about two centuries ago. With no species having much of an initial advantage, they have grown together, stretching upward in competitive unison.

The town of Jersey Shore is screened from the river by a maze of islands through which we came one afternoon. We ended up in the middle of the largest group of people we were to meet on the river, 20 students splashing about with nets. They were members of a freshwater ecology class organized four years ago by William Graff, a biology teacher at the Jersey Shore Area High School. At that moment Graff was hip deep in the West Branch encouraging his students as they measured current velocity, took samples of water and soil for laboratory testing and collected aquatic plants and animals. He says he has been very pleased with the elective course his students are enrolled in; he feels it has better acquainted them and their parents with the river that flows past their community.

Deb Waltz, a junior, had become thoroughly soaked dragging a net through the shallows. "I am collecting organisms," she explained.

"What kind?"

"Whatever organisms I can catch."

Waltz said that eventually she would like to be either a jockey or a veterinarian, but in the meantime freshwater ecology has been her best class. "I like anything alive, and I've found out about all sorts of things living in the river that I didn't know about," she said.

Each year, said Graff, his students are finding that the river has become more wholesome, less polluted and less acid. In consequence, life forms are more numerous and varied. Recently, for the first time in decades, trout rejoined the community of organisms in this stretch of the West Branch.

Since the mid-19th century, Williamsport has been the West Branch metropolis. It is now a city of 38,000. For a time after the Civil War it was the logging capital not only of the Susquehanna but of the nation.

Now Williamsport has turned to other industry and because of repeated flooding it has more or less turned its back on the river. Only marginal enterprises are still located directly on the river. These and the rest of the city are separated from it by high flood walls and levees. From canoe level, the effect is similar to running between canyon walls. It is hard to see much of Williamsport or get into it. Fortunately, we had only one bit of business to transact in the city—filling our water jugs. By and by we came to one of the less precipitous embankments, climbed it, crossed a railbed and descended through the green briers into the backyard of an elderly South Williamsport resident who was watering his garden. When we explained ourselves, he was very cooperative and enthusiastic. "Mother, Mother," he called to his wife. "Come out here. Here's a fellow and his daughter who have paddled a canoe all the way from above Clearfield. They're going to use our water."

"Clearfield," said the lady after she had come out of the house and heard the story for herself. "When we were in Florida we met a lovely couple from Clearfield. Their name was Hadley, or maybe Halsey. Do you know them?"

"Actually we live around Gettysburg," I said. "We just passed through Clearfield and didn't have a chance to talk to many people."

"You drove all the way up there to come down the river?" said the lady. "How fast do you go? Would you like some ice cubes in your jugs?"

"About 20 or 25 miles a day. Thanks, but the ice would just melt. All we need is something wet."

"You must have been in some wild country. Where did you get your water?"

"Up until now mostly from springs. Right out of the mountain. Just like old times."

Though we had feared the worst, Williamsport, to its credit, does not foul the river. The water remains clear and the acid has been all but dissipated. In fact, this lower section is the most vital in terms of aquatic life. One morning, from her vantage point in the bow, Lyn counted 103 of what she classified as "really big fish." Most of them were suckers and bass, perhaps 14 inches long.

Below Williamsport the major highways bypass the flood plain. Some country roads meander along the north bank, connecting villages that were built in the 18th century, prospered in the 19th and have long since become stable in size and expectations. The scene on that side is pastoral and antique. On the other bank it is still wild as the river beats against the last of the big Appalachian ridges—Bald Eagle Mountain. Surprisingly, the lower river still has some respectable white water. It is not created by pitch and speed as in the highlands, but by sheer power. As the current is forced through narrow interisland channels, some heavy standing waves are created. There is no need for maneuvering since these are straight, deep shots. It is a matter of getting set, more or less like riding a bucking horse.

All along the lower river are convincing signs of how strong and savage it is: the remains of a railroad bridge crumpled casually like a cheap toy, gaping cuts in the bank, the front quarters of a Chevrolet wedged high in a tangle of current-blasted oak. There are very good reasons for Williamsport and other communities to have stepped gingerly away from this river, and for fortifying themselves against it. Yet, at least for the visitor, if not for permanent residents, there is something grand about its power and latent savagery.

The West Branch is truly a wild river, though not of the picture-postcard variety. It has never been pampered, protected by legislative acts, or guarded by park rangers. We have had at it ferociously for the better part of two centuries. We have taken our best shots. We have used and abused it as we have few big rivers. It has been scarred and tainted by what we call civilization, but it has not succumbed. It has survived because of its great powers of resistance. It has held us at bay, defended its own wildness. To personify, perhaps outrageously, the West Branch is a river of great integrity, and that is why I have always admired it so much.

Boiling around Bald Eagle Mountain, still demanding that people and their works keep a respectful distance, it powers down to Northumberland and joins the Susquehanna, a more sluggish river, for the final run to the sea.

By chance we finished our trip on the eve of the stern paddler's 50th birthday.

"What a hell of a way to end the first 50 years," I said.

"And begin the second 50," Lyn said.

"Let's make a date for 2027," I said. "I may get hung up on a rock or a hard place, but you come back and check out the West Branch. You can tell me about it later."

"I'll be there," Lyn said.